The **Little Big Book** Series

Performance Acceleration Management (PAM)

Rapid Improvement to Your Key Performance Drivers

OTHER BOOKS PUBLISHED (OR SCHEDULED TO BE PUBLISHED) IN *THE LITTLE BIG BOOK* SERIES

Corporate Governance, H. James Harrington and Marcus E. J. Bertin, 2009

FAST

Fast Action Solution Technique, H. James Harrington, 2009

Value Proposition Identification, H. James Harrington and Brett Trusko

Initial Business Case Development, H. James Harrington and Frank Voehl

Organizational Project Portfolio Management, H. James Harrington and William Ruggles

Project Activities Management, Review, and Assessment, H. James Harrington

Performance Improvement Techniques and Sample Outputs, H. James Harrington and Charles Mignosa

Closing the Communication Gap, H. James Harrington and Robert Lewis

Using Software Technology to Improve Organizational Performance, H. James Harrington, Richard Harrington, Jr., and Ron Skeddle

Developing the Global Economy at the City, State, and National Levels, H. James Harrington and Abdul Rahman Awl

Creativity and Innovation throughout the Organization, H. James Harrington, Richard Harrington, Jr., and Ron Skeddle

Customer Cementing Relationships, H. James Harrington

The **Little Big Book** Series

Performance Acceleration Management (PAM)

Rapid Improvement to Your Key Performance Drivers

H. James Harrington

CRC Press
Taylor & Francis Group
Boca Raton London New York

CRC Press is an imprint of the
Taylor & Francis Group, an **informa** business

A PRODUCTIVITY PRESS BOOK

CRC Press
Taylor & Francis Group
6000 Broken Sound Parkway NW, Suite 300
Boca Raton, FL 33487-2742

Printed on acid-free paper
Version Date: 20130226

International Standard Book Number-13: 978-1-4665-7257-7 (Paperback)

Library of Congress Cataloging-in-Publication Data

Harrington, H. J. (H. James)
Performance acceleration management (PAM) : rapid improvement to your key performance drivers / H. James Harrington.
p. cm. -- (The little big book series ; 1)
Includes bibliographical references and index.
ISBN 978-1-4665-7257-7 (pbk.)
1. Organizational effectiveness. 2. Benchmarking (Management) 3. Quality control. I. Title.

HD58.9.H3728 2013
658.4'013--dc23 2012038880

Visit the Taylor & Francis Web site at
http://www.taylorandfrancis.com

and the CRC Press Web site at
http://www.crcpress.com

It is not how far you go, or the rewards and the recognition you receive that is important, but rather who walked with you in the path through life. I was fortunate to go hand-in-hand with my lovely wife Marguerite. This book and my life are dedicated to her.

Contents

Acknowledgments

I want to acknowledge Candy Rogers, who invested endless hours to edit and transform my rough draft into a finished product. I couldn't have done it without her help.

I would be remiss not to acknowledge the contributions made by my many clients to the broadening of my experience base and to the many new concepts that they introduced to me. Every engagement I worked on has been a learning experience that could not be duplicated in any university.

I'd also like to recognize my many good friends that have freely shared their ideas with me, helping me to grow and look at things in a new light. I would particularly like to point out the many contributions I received from Chuck Mignosa, Frank Voehl, and Armand Feigenbaum.

Preface

Plan for success or rush into failure.

H. James Harrington

"Every year we improve. I have data that proves this. Continuous improvement is a way of life in our company. Then why, oh why, do we continue to lose market share even though we are using the latest improvement methodologies and installing software like Customer Relationship Management (CRM)?" This is a typical comment that I have heard from organizations throughout Europe and North America. Middle-class, high-paying jobs have migrated out of Europe and North America, moving first to Japan and then on to China. Organizations that produce steel, automobiles, televisions, and now solar panels in the United States embraced the latest improvement approaches, like Six Sigma, process redesign, benchmarking and TQM (Total Quality Management), and had very positive results. They recorded major cost savings and quality improvements. Why is it then that they were not competitive with their Asian competitors? The answer is easy to understand. The U.S. organizations were improving, but not at a rate that made their comparative value to the consumer equal to or greater than the products manufactured in Asia (Figure 0.1).

Consumers buy products based on the perceived value that they get from the products. Value in the consumers' eyes is a combination of how they perceive the item's functionality, quality, reliability, and cost. Much of the quality and cost reduction resulted in increased profits for the organization, but this was not passed on to the consumer. Although the organizations within the United States were improving, their improvement rate was not adequate to make their outputs of equal- or better-perceived value in the eyes of the consumer. Today, there is a great deal of emphasis within the United States to bring back offshore production facilities to strengthen the American economy. However, these efforts will not be successful unless we greatly accelerate the performance improvement results within the organizations and facilities located within this country. The key to building a viable manufacturing base within the United States is to increase the quantity and capability of our skilled production working team and the support staff. In addition, competition around the world is

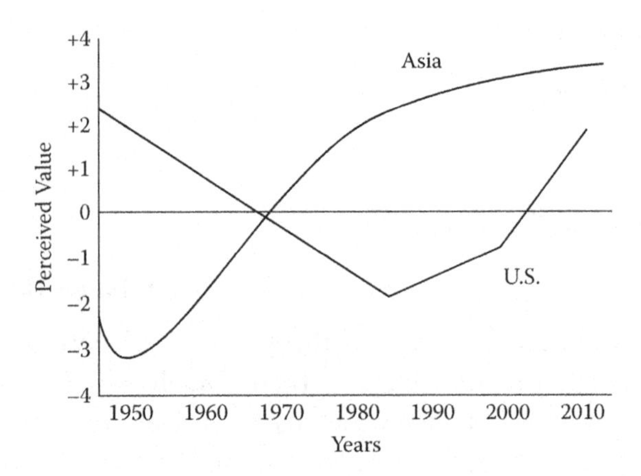

FIGURE 0.1
Perceived consumer value of American products versus Asian products.

increasing in the service industries. As a result, every organization, both service and manufacturing, needs to focus on providing more value to their customers than their competition can. This even applies to the government entities. Around the world, many government bodies and officials are being replaced because another entity is perceived by the general public as providing them with more valuable results. The problem with these organizations that are losing favor with the people they service is not that they are not improving, for in most cases, they are. It is that they are not improving fast enough to keep up with the expectations of the people they service. As a result, there is a worldwide need to accelerate the rate of change and improvement in order to meet the expectations of the people to which organizations provide products and service.

In the early 1950s, organizations in Japan realized that they were different from the organizations in the United States and, although they could benefit from the technologies developed in America, they also realized they were destined to remain a follower, not become a leader. Today the same common sense also is being reflected in American companies. They realize that they are different from Japanese companies and even different from other companies in North and South America. The leading companies are all realizing that to compete in the international market, they need to step back and stop reacting to the latest fads and stop copying other organizations. They are beginning to realize that it is essential that they look at how they want to change and to assess the benefits that all the different technologies and tools that are available to them can bring about a transformation. It takes time up front, but it saves total cycle time, cost, and effort over the life of the project. In addition, it produces much better results. If properly designed, it will create an organization that is creatively bringing out the best that each employee has to offer. We don't think this is true for just the big multinational companies; this approach is even more effective for the midsize and small companies. One excellent example of this is Globe Metallurgical (Beverly, Ohio), which was the first small business to win the coveted Malcolm Baldrige National Quality Award. Globe's continuous improvement plan consisted of 96 different objectives, with multiprojects' target completion dates distributed over a two-year period. Accurate (Flint, Michigan), a small 75-person company in the middle of the United States, is another excellent example of an organization that established a set of vision statements on how they wanted to change the organization and then developed the plan to make it happen.

I have worked for over 30 years helping organizations implement improvement approaches. We have used statistical process control, quality circles, teambuilding, balanced scorecard, design of experiments, process redesign, project management, Lean, total quality control, total quality management, total improvement management, nonverbal communications, employee surveys, customer surveys, total productivity management, CRM, streamlined process improvement, ERP (enterprise resource planning), Six Sigma, organizational change management, simulation modeling, strategic planning, organizational alignment, reengineering, continuous improvement, 5Ss, knowledge management, area activity analysis, innovation, and all the Japanese terms. I know of over 1,300 different approaches being sold to management to improve an organization's performance. There has been a constant flow of supposedly new tools offered that were often just old concepts that had been dusted off and given a new name. Every author and consultant develops his/her own twists to the basic concepts of good management in order to make a sale.

Each function within the organization has its own unique name for the tools it wants management to buy (Figure 0.2).

- The controller wants to install **Total Cost Management** with tools like activity-based costing that would reduce costs by 35%.
- The industrial engineering group pushes **Total Productivity Management** that they say will reduce costs by 35%.

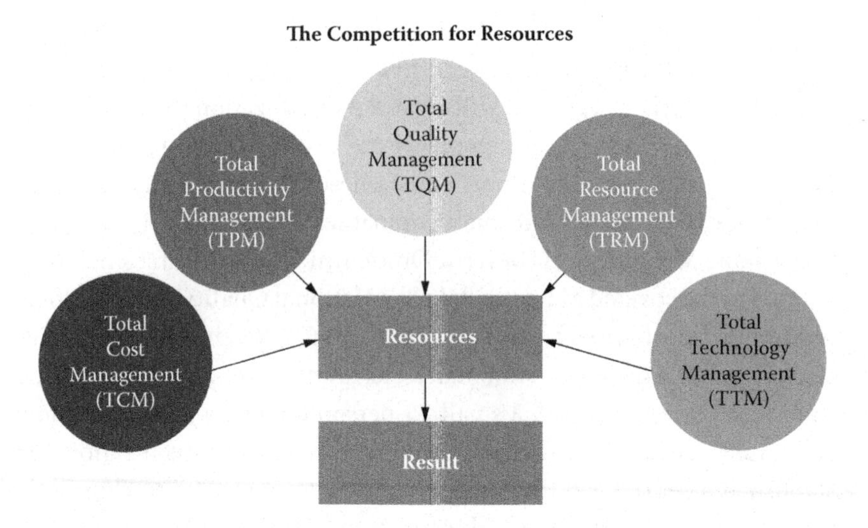

FIGURE 0.2
The competition for resources.

- The Human Resource (HR) department points out to management that its most valuable asset is its people and by focusing on improving the capabilities of its people through the use of **Total Resource Management** concepts, HR could reduce costs by 30%.
- The information technology group wants to install a **Total Technology Management** program. They explain to management that, in today's competitive environment, technology is the key driving factor and without the latest computers and computer programs, like CRM and ERP, there is no way the organization can compete with its competition. By investing in Total Technology Management concepts, they could reduce costs by at least 30%.
- And let's not forget the quality assurance group that pushes **Total Quality Management** as a single item that would greatly reduce costs, increase customer satisfaction, and allow the company to be competitive on the international market.

Note: By just investing in the first three initiatives, management eliminates 100% of the costs. By investing in all five approaches, management could be making a 30% profit and not have to bother with those customers that give them so many problems.

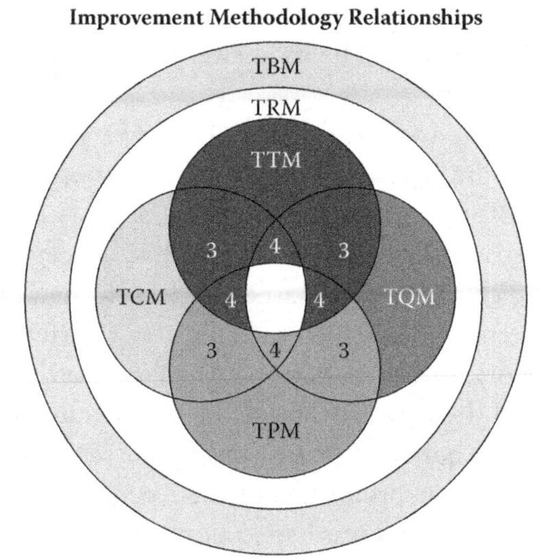

FIGURE 0.3
Improvement methodology relationships.

We all realize that you cannot eliminate all the costs. The reality of the situation is that each of the five approaches has a lot of overlap (Figure 0.3). They all focus on the use of teams, effective use of project management, and techniques, such as brainstorming. The truth of the matter is that management, on a whole, only has between 5 to 15% of the budget that can be used in discretionary spending. The rest of the budget is already directed to cover their basic operating costs, such as salaries, taxes, utilities, equipment, etc. All five approaches are competing to capture as much as possible of that 10 to 15% of the organization's budget that falls into the discretionary spending category. In addition, each of these five approaches require the executive team to be completely focused and dedicated to their specific approach ignoring all others. They also all require that the organization's culture be restructured in line with a specific approach for the projected returns to be realized. The fallacy in all of these performance improvement approaches has been that it is impossible for the executive team to only focus on one part of the business. In addition, none of them include an organizational structure that would bring about the required changes in the organization's culture that are needed to make the approaches permanent. As a result, management frantically shifts its attention from one approach to another resulting in short periods of ecstasy followed by long periods of despair and failure. Each of these approaches can have an initial positive effect upon the organization, but the enthusiasm related to the tools or approaches soon dissipates and management is off searching for a new silver bullet that would be its answer to the performance improvement requirement.

To make the situation worse, there has been a very high turnover in the executive team, so the experiences of the past are soon forgotten. This has provided an excellent opportunity for many authors of new books and consultants to repackage these old but effective concepts. By modifying them slightly and changing their names, it gives the new executive team a supposedly different tool in which to bring about performance improvement within the organization. Add to it the fact that the expectations of the customers and the performance of the competition are continuously improving, which requires all organizations to set higher performance requirements upon their products and their people. For example, when CRM was first introduced, it was hailed as an approach that would greatly improve your competitive position by establishing better customer relations than your competition was able to obtain. As a result, organizations poured hundreds of thousands of dollars into installing CRM systems and training their people to use them. Unfortunately, the improvement in competitive positioning in most cases was not realized because the competition was also installing CRM systems. As a result, installing a CRM system did not provide a competitive advantage to most organizations, but not installing one proved to be a very negative disadvantage. The end result was that most organizations were required to install CRM systems that cost hundreds of thousands of dollars without an increase in competitive advantage because the standard for customer service had been reset at a much higher level.

The real problem that we have with what we have been doing is we have been focusing on changing the tasks that we perform rather than on the behavior of our executives and our employees. It's only when we bring about profound changes in the behavioral patterns and beliefs of the executives and the employees that we can have an effect on the culture within our organization. It's relatively easy to bring about short-term changes within the organization, but to embed them permanently into the behavioral patterns and culture of the organization is very difficult and requires different fundamental approaches than most organizations have been using today. We need to define the Key Performance Drivers (also called "key controllable factors") that drive the behavioral patterns and culture of our organization because they drive the organization's long-term performance. The key to long-term accelerated performance improvement is defining these

Key Performance Drivers (KPDs) by analyzing them to determine how they need to change to meet the challenges that lie ahead of the organization. Once you have defined what you need to change, you then can select the tools that will address the problems that each KPD is facing in today's environment and the roadblocks that are required to transform it from today's AS IS (Current State) condition to the desired future state. Only in this manner can an organization hope to bring about the cultural changes that are required to survive in this rapidly changing environment. This is true of both our public and private sector organizations.

There is no doubt about it. Most organizations throughout the United States and Europe have managed to implement effective continuous improvement initiatives. Unfortunately, these continuous improvement initiatives have been short-lived and often poorly directed. The result has been that the magnitude of these improvements was not great enough to close the perceived value gap between the products manufactured in Europe and the United States. In order for European or American organizations to provide products and/or services that can meet the value standards set by Asian companies, the U.S. and European organizations and governments need to greatly accelerate and restructure their improvement efforts in order to be competitive on the international market and to offset the negative trade balance that is being created by European and American countries with the Asian countries.

The many seemingly different programs we have been using over the past 30 years have improved many organizations' performance improvement progress. We started using quality circles with the focus on team building. We then migrated to Total Quality Control with a focus on minimizing the defects as delivered to the external customer. The next major initiative was Business Process Improvement. In this case, the focus was on improving the efficiency, effectiveness, and adaptability of the support and service areas. The mid-1980s introduced the ISO 9000 quality standard whose theme was to document what you do and do what you document. This forced many organizations to introduce a great deal more structure into their operations. Then along comes Six Sigma with a focus on lowering the error-opportunity level of no more than 3.6 errors per million opportunities. Of course, this led to individual managers looking for different ways to count the number of opportunities for errors. Accompanying the Six Sigma initiatives was the development of a separate group of elite problem solvers called Black Belts and Master Black Belts. In these cases, continued existence for the individual Black Belt was tied to cost reduction. About this time, the concept of a Balanced Scorecard came into play and a whole different set of measurements took on management's attention. Along with this new attention on key business measurements, customer satisfaction became a major focus within most organizations.

In the early 1990s, the Process Reengineering methodologies became the "in thing" for many organizations. It focused on the use of technology to bring about step-function-improvement in the organization's performance. Based on Toyota's success in capturing the worldwide automotive market, Toyota's manufacturing process became the model for many organizations. Starting approximately in 2010, the "in thing" for everyone who was interested in performance improvement was a program called Lean. The focus of Lean was to eliminate all waste from all parts of the organization. Key focus in a Lean operation started with elimination of clutter and an organized clean work environment. Unfortunately, some executive teams quickly identified waste involved in paying higher wages to workers in Europe and North America when they could get the same job done in Asia at a small percentage of the cost.

> I love and respect the gurus of quality. It is just too bad they couldn't agree on how to create quality.
>
> **H. James Harrington**

Is it any wonder that management is confused? Even the individuals who were recognized as the gurus in the continuous improvement processes could not agree on how an organization should implement the improvement process.

Philip B. Crosby's (author and practitioner of quality management) "14 steps" focused on motivating individuals, documenting their commitment to quality by having them sign pledge cards and measuring progress through the use of quality costs. (This was a concept developed by Armand V. Feigenbaum (quality control expert) in the 1950s.) The 14 steps of quality improvement include

1. Management commitment
2. Quality improvement teams
3. Measurement
4. Cost of quality
5. Quality awareness

6. Corrective action
7. Zero defects planning
8. Employee education
9. Zero defect day
10. Goal setting
11. Error-cause removal
12. Recognition
13. Quality councils
14. Do it over again

Dr. W. Edwards Deming introduced Japanese top management to the statistical process control methods developed by Walter Shewhart in the 1920s. Japanese management was quick to recognize that this was the "secret weapon" that allowed the United States to mass produce the vast quantities of high-quality weapons that defeated Japan in WW II. Dr. Deming developed a different "14 point" program just for the United States.

Just before he passed, Dr. Deming began to advocate a system he called "Profound Knowledge" that is made up of another 14 points. They include:

1. Nature of variation
2. Losses due to tampering (making changes without knowledge of special and common causes of variation)
3. Minimizing the risk from the above two (through the use of control charts)
4. Interaction of forces, dependencies, and interdependencies
5. Losses from management decisions made in the absence of knowledge of variation
6. Losses from the successive application of random forces that may be individually unimportant (such as workers training other workers)
7. Losses from competition for market share and trade barriers
8. Theory of extreme values
9. Statistical theory of failure
10. Theory of knowledge in general
11. Psychology, including intrinsic and extrinsic motivation
12. Learning theory
13. Need for the transformation of leadership from grading and ranking
14. Psychology of change

Dr. Feigenbaum focused his efforts on 10 benchmarks that direct the improvement effort. His 10 Benchmarks for Quality Success are

1. Quality is a company-wide process.
2. Quality is what the customer says it is.
3. Quality and cost are the sum, not a difference.
4. Quality requires both individual and team zealotry.
5. Quality is a way of management.
6. Quality and innovation are mutually dependent.
7. Quality is an ethic.
8. Quality requires continuous improvement.
9. Quality is the most effective, least capital intensive route to productivity.
10. Quality is implemented with a total system connected with customers and suppliers.

Dr. Feigenbaum is the father of Total Quality Control and published the first book on the subject in 1951 (*Quality Control: Principle, Practice, and Administration*. New York: McGraw-Hill). He also originated the concept of Quality Costs. He looks at the total product value cycle and applies systems engineering approaches to bring about improvement.

Dr. Joseph M. Juran (often called the "father" of quality), on the other hand, fosters the belief that an improvement effort is driven by many small, step-by-step improvements. Each saves the company approximately $100,000. He uses Pareto analysis, which he created, to define the critical few problems and assigns teams to solve these problems. Dr. Juran defines quality as "fitness-for-use." He looks at what he calls "The Spiral of Progress in Quality." The quality function is the entire collection of activities through which we achieve fitness-for-use, no matter where these activities are performed. It includes:

1. Market research
2. Product development
3. Product design/specification
4. Purchasing/suppliers
5. Manufacturing planning
6. Production and process control
7. Inspection and test
8. Marketing
9. Customer service

Dr. Kaoru Ishikawa was the leading quality expert from Japan and the originator of the quality circle concept. He espoused that the best way to improve performance was through the empowerment and enlightenment of the employees. Dr. Ishikawa's concepts fueled the unparalleled expansion in employee team skills and problem-solving training. Although Dr. Deming and Dr. Juran are given credit for the miraculous transformation of Japan, Inc. (the production system that was developed by Japan after World War II), I believe that Dr. Ishikawa was the real genius because he took many concepts, put them together, and implemented them all effectively. Without Dr. Ishikawa's activity, I believe the work of Deming, Feigenbaum, and Juran would have had little effect on the Japanese.

Dr. Ishikawa looked at quality as a way of managing the total organization. He saw the management transformation as six categories. They were

1. Quality first—not short-term profits.
2. Consumer orientation—not producer orientation. Think from the standpoint of the other party.
3. The next process is your customer—break down the barrier of sectionalism.
4. Using facts and data to make presentations—utilization of statistical methods.
5. Respect for humanity as a management philosophy—full participatory management.
6. Cross-function management.

To add to all this confusion, there was the quality professionals' fixation on training management in how to speak Japanese. Programs and terms like *kaizen, kaikaku, Kaizen Blitz, heijunka, gemba, hoshin kanri, jidoka, kanban, muda, mura, muri, poka-yoke, sensei*, etc., were added to the performance improvement vocabulary primarily to fool management into believing they were being presented with something new.

Is it any wonder that the employees became confused and developed an attitude of "we will just wait this one out and it too will pass." Again this confusion existed and large sums of money and effort had been wasted because no one had taken the time to crisply define how the organization's culture needs to change in order to accelerate the performance improvement results that are required to catch up and surpass the international competition that in many cases has the distinct advantage of lower labor costs and less environmental restrictions. To eliminate this confusion and waste of effort and money, a methodology called Performance Acceleration Management (PAM) has been developed. This methodology uses the organization's KPDs to provide a vision of how the organization needs to change and a five-year plan to direct how the organization will accomplish this change.

> If you don't change behaviors/habits, the organization's culture hasn't changed. If you don't change culture, you won't have long-term performance improvement.
>
> **H. James Harrington**

1

Performance Acceleration Management: Its Theory and Practice

Most organizations are improving, but they are not winning the race.

H. James Harrington

INTRODUCTION

Everyone is talking about the need for a cultural change, but I believe that focusing on it is the wrong answer to today's problems. It doesn't prepare most organizations to prosper in the twenty-first century. Culture is defined as one's background, history, heritage, religion, your beliefs. Most organizations want to hold onto their culture and, in fact, are worried about losing it. Americans should be proud of the culture they have worked so hard to create. It is a culture rich in imagination, hard work, caring, risk taking, and accomplishments. It is a culture that has made the United States the richest, the most powerful, most productive nation in the world. Our culture is not the problem. It is the personality of today's population that is the problem. We talk about "workaholics" like "work" is the worst four-letter word in the English language. People work overtime begrudgingly if they are notified 72 hours in advance, and, if not, they refuse. It is the personality of today's workforce and our children that needs to be changed. Personality is defined as an individual's or group's impact on other individuals or groups. We need to change the personality of our people before we lose the culture that our forefathers worked so hard to create. It is the personality of our key managers that dictate the personality of the total organization. When a new CEO (chief executive officer) is appointed, the total organization adapts to his or her personality. If he or she is a baseball fan, you would be surprised how many people all of a sudden know last night's baseball scores. Just picture the organization's culture as a rubber band stretched around four posts to form a rectangle (Figure 1.1). When a new CEO/president takes charge of the organization, the rectangle is distorted forming a temporary culture to meet the personality of the individual (Figure 1.2). As soon as the CEO/president moves onto another assignment, the rubber band snaps back to its original rectangular shape as seen in Figure 1.1, because this is a culture of the organization. Often the next new CEO/president's personality and hot points cause the temporary culture of the organization to be shifted again in line with the new leader's personality and preferences (Figure 1.3).

We cannot go back to what we used to do because this old world has changed. The amount of information available to the individual doubles every two years, according to the annual Digital Universe study consultancy from IDC.[1] The daily edition of *The New York Times* contains more

FIGURE 1.1
The organization's culture before new management.

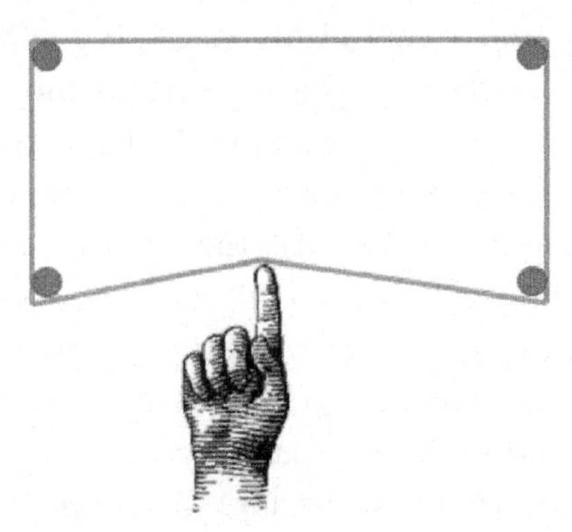

FIGURE 1.2
The effect of new management on the organization's temporary culture.

FIGURE 1.3
The effect of a second new president on the organization's temporary culture.

information than the average person was likely to come across in a lifetime in seventeenth-century England.

Let's focus on what we can do to influence and change the environment that impacts the personality of today's workforce. What the organization needs is to develop a plan that will change the environmental factors that impact the personality of the employees, placing special emphasis on the management team. If we sustain a positive change in the personality of the organization for a long enough time (about five years), we would change the organization's culture.

> We can have more than we've got because we can become more than we are.[2]
>
> **Jim Rohn**
> *Excerpts from The Treasury of Quotes*[2]

PERFORMANCE IMPROVEMENT PLAN VERSUS BUSINESS PLAN

> If you change the way the outside world views the organization, it is just a temporary fix. It becomes permanent when you change the way the organization views itself (its culture).
>
> **H. James Harrington**

There is a big difference between a performance improvement plan and a business plan. The business plan sets the product and service strategy for the organization: the market that they will hope to penetrate, the new products that will be introduced, the production strategy, etc. It is a plan that directs and guides the business. The business plan is primarily directed at meeting the needs of only two of the stakeholders associated with the organization: the customer and the stockholder. It is a plan that is primarily focused on the external opportunities. On the other hand, the performance improvement plan is an internally focused plan that is designed to transform the environment within the organization, to change its personality (behavioral characteristics). It takes into consideration the needs of all the organization's stakeholders from an improvement standpoint. The performance improvement plan defines the transformation in the business personality of the organization. It provides an orderly passage from one state or condition to another. The performance improvement plan supports the business plan so the two of them, although different in content and intent, must be kept in close harmony.

Why Do You Need a Detailed Plan to Manage Performance Improvement?

Now you may be thinking: "Why does my organization need to develop a performance improvement plan to improve? I know a lot of problems that we can start working on right now." You cannot afford to stop putting out the fires that your present organization is fueling, but as long as you continue to do what you have been doing, you will continue to get the results that you have been getting. Unfortunately, probably your competition is not content with their situation and they are changing, so, if you don't change, they will improve their competitive position. In fact, the twenty-first century will be dominated by those organizations that improve the most and can change the fastest.

Most organizations, in their hunger to improve their relative performance, have embraced many different improvement tools. It seems like each time someone went to another conference they came back with another improvement tool. Our research shows that more than 1,300 different improvement tools exist today that will provide a positive impact on an organization's environment. Each of these tools work under the right conditions. Many solve the same type of problems; others are the same tools with a different name. The following is a list of 10 of the many tools that could be used to improve an organization's management leadership and support.

1. Management self-audits
2. New performance standards
3. Improvement policy

4. Improvement visions
5. Annual strategic improvement plans
6. Leadership skills development
7. Self-managed work team
8. Responsibility charting
9. Honesty and integrity
10. Risk/opportunity management

No organization can afford or effectively utilize all of the 1,300 improvement tools. Many of the tools overlap in approach and the problems that they solve. Many of them are not applicable or have little impact upon your organization.

Just as individuals differ, organizations differ in many ways as well. They have different management personalities, customers, products, cultures, locations, profits, quality and productivity levels, and technology and core capabilities. Added to this complexity is the fact that winning, surviving, and losing organizations have to do a very different set of things to improve. It becomes readily apparent that there is no one approach to improvement that is correct for all organizations or even for different locations within an organization.

What Factors Affect an Organization's Culture?

> Wisdom means keeping a sense of the fallibility of all our views and opinions, and of the uncertainty and instability of the things we most count on.
>
> **Gerald Brenan**

The organization's culture is created over a long period of time as a result of the way management implements the organization's basic beliefs combined with the way the employees react to management's stimulation.

Definition

Value: The basic beliefs that the organization is founded upon. The principles that make up the organization's culture are often called *values*. These are prepared by top management. They are rarely changed because they must be statements that the stakeholders hold and depend on as being sacred to the organization.

There is a great deal of confusion about which term or terms to use: value, beliefs, or principles. I don't care what you call them, but every organization should have a set of statements that communicate to management and to the employees an understanding of what the organization's culture is based on. These statements provide direction to management that governs their performance. To the employees, they provide a promise of conditions that the organization is built upon. Call them basic beliefs, guiding principles, or operating rules. Call them what you will. The important thing is that they must be defined, and the organization must live up to them. They are what I call the "Stakeholders Bill of Rights." See the following examples.

- Ford Motor Company's guiding principles are:
  - Quality comes first.
  - Customers are the focus of everything we do.
  - Continuous improvement is essential to our success.
  - Employee involvement is our way of life.
  - Dealers and suppliers are our partners.
  - Integrity is never compromised.
- The following is a quotation from IBM's Manager's Manual:
  - "An organization, like an individual, must be built on a bedrock of sound beliefs if it is to survive and succeed. It must stand by these policies in conducting its business. Every manager must live by these policies when making decisions and in taking action."
- The U.S. government has had a set of basic beliefs since the beginning. They are called the Bill of Rights and they have helped guide the United States for more than 200 years.

These fundamental statements, which we call basic beliefs or values, are the things that attract new employees to your organization. They define the rules on which the organization will not compromise. They are the employees' Bill of Rights, so they shouldn't be changed frequently and then only when they have become obsolete because of social and/or external environment. I worked for IBM for 40 years and their basic beliefs remained constant throughout that period.

Management's job is to live every business minute complying with the organization's values. This is the first obligation of every manager. No employees should accept a management position unless they have already lived up to these values and no manager should be left in a management role unless he/she lives up to the organization's values. Employees have a responsibility to work in support of these basic values and refusing to compromise on them. Management that asks employees to bend these basic values is not doing its job, and this condition should be brought to the attention of upper management.

What Are the Key Performance Drivers?

> You must define what and how you want to change before you can select the tools that will bring about the change.
>
> **H. James Harrington**

Management has control over relatively few things. They do not control the economy of the nation, their customers, their competitors, their suppliers, government regulations, the stock market, and so on. The organization is impacted by many things that management cannot control. Management cannot control what their competitors are going to do, government regulations, fluctuations in the stock market, the economy, etc. The only things that management can change are the Key Performance Drivers (KPDs). If management wants to bring about a change in the organization, they must change the KPDs within the organization because these impact the desired results. These KPDs are the things that the organization can choose to invest or not invest money or effort into so as to bring about changes in the organization's performance. (For example, they can invest money and resources in providing additional training and skills to the employees in order to improve organizational performance or they can invest effort and money in working with suppliers so that they can improve the quality and timeliness of the products that the organization depends on to satisfy the organization's external customers. The organization can decide to modernize the plant by bringing in new equipment and robots to minimize the manual labor that is required to produce a product.) These are typical examples of KPDs that the management team can invest money and/or effort in in order to improve the performance of the organization. The only way any organization can improve is by investing in these KPDs and determine how much to invest in each in order to stay ahead of the competition and their customers' expectations.

It may be that the future HP hero will not be limited to the engineer with the new idea for a circuit, but will also include the person who can make groups of people separated by product type and geography work together successfully.

Lew Platt
Executive vice president
Computer Products Sector

THE PERFORMANCE ACCELERATION MANAGEMENT APPROACH

Performance Acceleration Management (PAM) is a methodology directed at bringing about significant change in the organization's long-term performance. It is a methodology that brings about major transformation in the organization's culture. It is made up of seven phases (Figure 1.4.). They include:

- Phase I: Conducting an improvement requirements' assessment
- Phase II: Developing vision statements
- Phase III: Defining desired behavioral patterns and performance goals
- Phase IV: Developing individual KPD transformation plans
- Phase V: Developing a five-year combined PAM plan
- Phase VI: Implementing the combined PAM plan
- Phase VII: Continuously improving

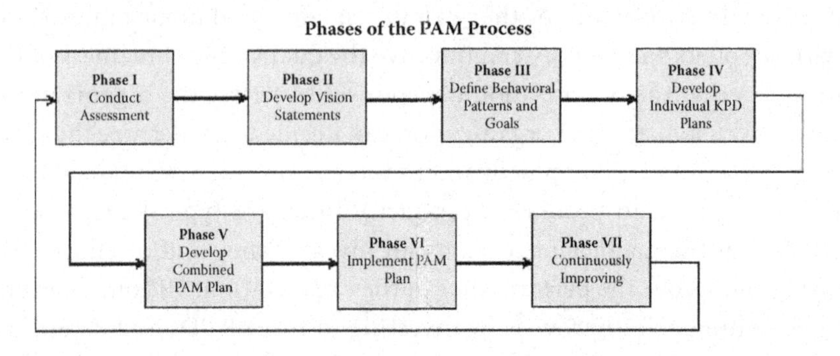

FIGURE 1.4
The seven phases of the PAM methodology.

Phase I: Conducting an Improvement Requirements Assessment

All projects need to start by establishing a firm documented base that will be used to direct the project. Managing performance improvement is no exception to this rule. There are two activities that take place during this phase:

- Activity One: Organize for performance acceleration
- Activity Two: Define present status and improvement opportunities

Phase II: Developing Vision Statements

Definition

Vision Statement: A vision statement is a documented view of the future desired state condition. A vision statement should be a short paragraph no more than three sentences long (a one-sentence vision statement is preferable). A vision statement should stretch the organization to be the best that it can be. Documented vision statements provide an effective tool to help develop objectives and improvement programs.

During this phase of the PAM process, the organization will develop a separate vision statement for each of the KPDs. These vision statements will reflect the desired operational status of the KPDs three years in the future. Phase II consists of eight activities:

- Activity One: Review and upgrade if necessary the organization's mission statement, values, and vision
- Activity Two: Review the assessment report prepared during Phase I
- Activity Three: Define the AS IS description for each KPD
- Activity Four: Develop preliminary KPD vision statements
- Activity Five: Conduct focus groups
- Activity Six: Conduct and analyze an Organizational Change Management survey
- Activity Seven: Prepare final KPDs' vision statements
- Activity Eight: Communicate the final KPDs' vision statements

Phase III: Defining Desired Behavioral Patterns and Performance Goals

There are two distinct and separate parts to Phase III: (1) defining the desired behavioral pattern changes that will occur within the organization and (2) defining goals for improvement in key measurements related to the different stakeholders within the organization.

Defining Desired Behavioral Patterns

During this part of Phase III, the PAM Steering Committee will review each of the vision statements to define the desired behavioral patterns that should be prevalent in management and the employees' behaviors. (For simplicity sake, the PAM Steering Committee will simply be referred to as the Steering Committee throughout the rest of this book.) The Steering Committee then prepares a list of these desired organizational behaviors. It is important to note that changes in behavioral patterns can be observed prior to improvements in key measurements, such as customer satisfaction and reduced cycle time. This is also the correct point in time to modify the rewards and recognition system to reflect the desired changes in behavioral patterns. It is very important to put in place early in the PAM methodology a reward and recognition system that reflects changes in the desired behavioral patterns. This part of Phase III consists of three activities:

- Activity One: Defining the desired behaviors/habits related to the vision statements
- Activity Two: Defining the desired and undesired behaviors/habits related to the activities as performed within the organization
- Activity Three: Defining how to measure desired behavioral patterns

Defining Performance Goals

During this part of Phase III, the Steering Committee will establish yearly high-level performance goals that are designed to make the organization more competitive than its competition. These goals are designed to accelerate the organization's performance over the next five years. They are used to prioritize the performance improvement initiatives so that they focus on the key performance measurements of the total organization.

This part of Phase III consists of the next five activities making a total of eight activities that are included in Phase III. They include:

- Activity Four: Defining key performance measurements
- Activity Five: Defining present performance levels of the key performance measurements
- Activity Six: Defining competitors' present performance level for the key performance measurements
- Activity Seven: Projecting competitors' performance level for the key performance measurements
- Activity Eight: Finalizing the key performance goals for the next five years

Phase IV: Developing Individual KPD Transformation Plans

During Phase IV, the Steering Committee will assign each of the vision statements to a subcommittee that will be responsible for preparing the individual KPD transformation plan for the assigned vision statement. This plan will define the tools/methodologies that will be used to bring about the required transformation as well as a projected timeline chart for the implementation of the tools/methodology. For this phase, the timeline chart will only state the number of months/weeks that is required for the various activities in implementing the tool/methodology. It also should suggest an individual or department that should be assigned the responsibility for coordinating the activities required to implement the tools/methodology. Phase IV consists of six activities:

- Activity One: Assigning a planning team (subcommittee) to each KPD vision statement to develop an individual KPD transformation plan
- Activity Two: Defining present-day problems
- Activity Three: Defining roadblocks to evolving to the desired future-state vision
- Activity Four: Selecting tools/methodologies to address defined problems and roadblocks
- Activity Five: Developing an implementation timeline chart for each tool/methodology
- Activity Six: Obtaining approval for the individual KPD transformation plans

Phase V: Developing a Five-Year Combined PAM Plan

> If I had 60 minutes to save the world, I would spend 55 of those minutes defining the problem and the other five solving the problem.
>
> **Albert Einstein**

During Phase V, the Steering Committee, in conjunction with the subcommittees for each KPD transformation plan, will meet to establish a timeline chart that will combine the tools/methodologies required to bring about the total performance acceleration activities for all the vision statements. There are a number of important factors that need to be considered as the individual tools/methodologies are scheduled into the organization's activities. Some of these are typical considerations, such as considering the other projects that are going on within the organization, the timing related to the goals in the key performance indicators, the workload of the areas impacted by implementing the tools/methodology, fluctuations in workload at various times of the year, and skills available to implement the tools/methodologies. Phase V consists of five activities:

- Activity One: Defining resource constraints
- Activity Two: Defining interrelated tools/methodologies
- Activity Three: Prioritizing individual tools/methodologies
- Activity Four: Preparing a Combined Work Breakdown Structure (WBS) for the PAM project
- Activity Five: Assigning individuals or departments that will be responsible for the successful implementation of each of the tools/methodologies

Phase VI: Implementing the Combined PAM Plan

During Phase VI, the combined PAM plan will be implemented. Each department responsible for a tool/methodology will prepare a detailed implementation plan based on the schedule that was defined in the WBS developed in Phase V. These detailed plans will be integrated together into a combined PAM plan. This plan will include critical tollgates that will ensure the organization that the project is on schedule and that the expected results and goals will be met. Phase VI consists of six activities:

- Activity One: Developing individual detailed implementation plans for each tool/methodology
- Activity Two: Combining the individual detailed implementation plans into a Rolling 90-day WBS
- Activity Three: Preparing a three-year financial plan to fund the PAM project
- Activity Four: Establishing the tracking system to ensure the project is on schedule, within costs, and will produce the desired results
- Activity Five: Establishing a measurement system that would measure the impact the project is having on the organization's performance
- Activity Six: Evaluating contributions made by individuals, groups, and teams, and recognize outstanding performance

Phase VII: Continuously Improving

As individual projects/activities are completed and the implementation teams are reassigned, procedures need to be put in place to ensure that the gains that were realized can be maintained. Too often, once an organization reaches a specific goal or implements a specific tool, the organization relaxes with the belief that the job is done. This is a major mistake that many organizations make. When organizations stop improving, they are not holding their own, they are slipping backwards because their competition is probably continuously improving. No organization, no matter how good or how high they are performing, can afford to relax because those organizations set the benchmarks and they are the targets for all their competitors. Too often, once the spotlight is taken off of an individual activity, it starts to decay. It is for these very reasons that so many of the performance improvement initiatives that have been undertaken in the past 50 years are considered failures by many executives. To truly excel, we have to have an inbred feeling that we are not as good as we can be, that there is always further room for improvement. All too often we don't have time to do it right every time, but we often have to find time to do it over again when it goes wrong. In truth, there is no time in life for "do overs."

SUMMARY

Short-term success within an organization can often be accomplished by the use of specific tools or methodologies. But, in the long run, it is the culture of the organization that is going to make the real difference between success and failure. Organizations throughout the world have made a major mistake in trying to bring about major performance improvement within their organization by implementing the latest fad without considering how they need to change the culture within the organization. Today, we live in a very complex environment and many different factors impact the performance of any organization. This complexity makes it mandatory that the organization step back and do some real soul-searching to define how it wants to position itself in the future. This means that we must define how and what we want to change before we select tools/methodologies that would bring about changes in the way the organization functions. Organizations have realized for some time that there is a need for a strategic plan related to their products and how they are viewed from the external sources. Organizations around the world are just beginning to realize that they need to put as much effort into developing a strategic plan that looks at the organization's internal operations and culture. This new internal-focused strategic plan will define how the PAM methodology will be implemented.

> Taking time up front to plan it right reduces the cost and cycle time of the total project.
>
> **H. James Harrington**

REFERENCES

1. IntelliBriefs. 2011. The amount of data doubles every two years. Digital Universe study, fifth edition. www.clusterofthoughts.com/content/amount-data-doubles-every-two-years. html.
2. Rohn, J. 1994. *The treasury of quotes.* Lake Dallas, TX: Success Books.

2

Phase I: Conducting an Improvement Requirements Assessment

> If you don't know where you started from, how can you measure progress?
>
> **H. James Harrington**

INTRODUCTION

Phase I is designed to acquaint the organization with the Performance Acceleration Management (PAM) approach to accelerated performance improvement. To accomplish this, most organizations will assign a high-level Steering Committee to gain understanding of the approaches and to conduct an assessment to determine its application to the needs of the organization (Figure 2.1). Making the decision to apply the PAM methodology in your organization is a major decision that will have a long-term impact on the performance and culture of the organization. Making this commitment must not be given in a haphazard manner because the approach leads to the development of a combined five-year PAM plan that will transform the organization's culture, reputation, and performance. There are a total of two different activities that make up Phase I:

1. Activity one: Organize for performance acceleration
2. Activity two: Define present status and improvement opportunities

Defining How to Change

Start with an organizational assessment to determine the most critical change areas to focus on first.

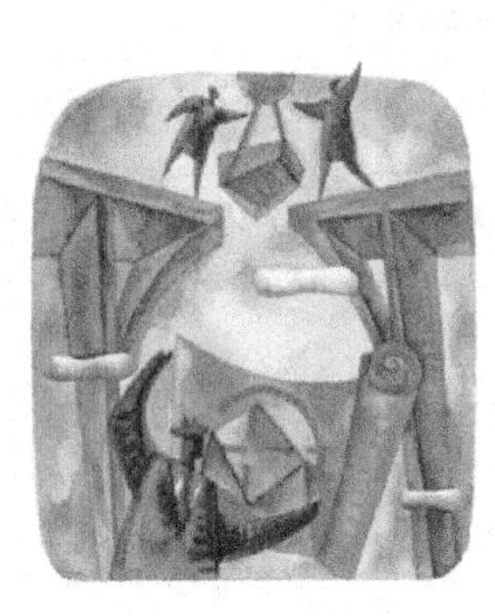

FIGURE 2.1
Defining how to change.

Activity One: Organize for Performance Acceleration

To be realistic, before an organization starts Phase I, someone in the organization had to become interested in the PAM methodology. This usually occurs because an individual has read a book, attended a conference, or worked for another organization that had already implemented the approach. This individual, in turn, must convince key people on the executive team that the organization should, at a minimum, enter into Phase I of the PAM approach to accelerate performance improvement. This individual plays a key role in sponsoring and acting as the champion for the PAM project.

A key element in the PAM methodology is a well-thought-out creative design that is supported with an ambitious, but realistic, implementation strategy. The group responsible for the design and its implementation is the PAM Steering Committee (PAMSC). Throughout the process, this Steering Committee will oversee and provide guidance and direction to all activities that are related to the PAM initiative. This is a high-level committee and its members need to be empowered to make critical business decisions. As a result, the senior member of each function within the organization (usually a vice president), along with the COO/president of the organization, will be selected to join this Steering Committee. In addition, the individual who will champion the PAM project and the key technical leaders of the organization should be included, such as the Human

Resources (HR) manager, the Project Office manager, the Information Technology (IT) manager, and a representative of the union, if it is a union shop. Typically, an individual will be assigned to serve as a facilitator of this committee.

To start this phase of the process, the CEO or the COO should call top executives and union office officials together to discuss the need for accelerated performance improvement and what action should be taken to accomplish this. Such a meeting could go something like this:

> The COO of Johnson Plastic, Inc., John E. Johnson, starts the meeting by saying, "Ladies and gentlemen I am concerned about the slow rate that the organization is improving its performance. Although we have sustained a continuous improvement in our level of performance, we have not been closing the gap on our competition. I have talked to each of you about this individually, so I know most of you share my concerns. I am sure that the very effective rumor mill has already informed you that I have asked Frank Parker to take on a special assignment on my staff to implement a performance acceleration initiative throughout the organization, and that includes your sales organization, Mr. Weston."
>
> Ray Weston, the vice president of sales and marketing for the organization, had been jotting down some notes about the meeting he had just left to come to this one. Ray quickly looks up, closes his notebook, and says, "I'm with you, John."
>
> Johnson continues, "I suppose all of you have been waiting for the second shoe to fall, and here it is. We are good and I'm proud of the accomplishments we have made over the years. I believe that we have one of the best teams in our type or any other type of business, but—and that's a big 'but'—we can be much better than we are. Our profits are high, our order backlog is good, and our sales prices are in line with our competition, but it is costing us a lot to put out our products. We should be able to boost our profits by over 90% if we can stop making so many needless errors. At about half the meetings I hold, someone reports that things did not get done on schedule because something wasn't considered or because someone didn't get the information they needed or someone provided the wrong data or the activity is taking longer than was originally estimated. Just six months ago, marketing estimated that the market demands for the next 12 months for our 920 units would be 10,000 units per month. Now sales is telling me that they could sell 25,000 a month if they had them."
>
> Ray Weston quickly speaks up, "That unit took off like a skyrocket. No one could have predicted that the customer acceptance level would be that high."
>
> Johnson says, "Ray, I'm not trying to pick on you. I could give similar examples of errors from every function in this room and that includes me.

Some of them have had long-term financial impact on our organization, and others have been just minor inconveniences. The problem we have is that we are beginning to accept errors as the normal way of doing business. The management personality of our company has been degraded to the point that we are accepting errors and mistakes as a way of life; instead we should all be upset and irate when they occur. Most of us have developed two standards, one for our personal life and one for our professional life. In our personal life, we expect everything to go right and when it doesn't, we become irate. For example, if your doctor treated you for pneumonia and you had an appendicitis attack, you wouldn't stand for it. You would first tell him what you thought of his ineptness and then you would call your lawyer. However, at work, if something doesn't go according to schedule or someone on the team causes an error, you reason that everyone can make a mistake, so you expect errors and plan for them. Instead, each of you should get mad when you see the waste, scrap, and rework around the company, just as you would if accounting left the first digit off your paycheck. We cannot accept any level of error to be considered normal for this organization. This will mean that we all need to step back and look at the way we are behaving. As leaders in this organization, we need to not only make good decisions, but we need to behave in a manner that sets the example for our employees. Benchmarking our competition and staying in line with them is no longer good enough. We have to accelerate our performance improvement to the point that we provide more value to our customers than our competition can. I realize that much of our competition is outsourcing their production operations, thereby gaining a significant reduction in labor costs and allowing them to reduce their sales price. This is widening the gap in the value that their product has in the eyes of the consumer as compared to our products. It is our responsibility, as leaders in this organization, to accelerate our performance improvement activities so that the product manufactured here in the United States still represents greater value to the consumer than the products manufactured overseas. This is the obligation we have to the people that invested in our organization and to our employees. So, in order to get the performance acceleration initiative started, I am forming a Performance Acceleration Steering Committee, and I expect each of you to actively, personally participate in this committee's activities. This committee will be chaired by Frank Parker. The three-fold mission of this committee is to:

1. Develop an organization-wide strategy to implement the performance acceleration process
2. Provide direction and guidance to the implementation
3. Adjust the process to meet changing business needs"

Mary Cross interrupted saying, "That sounds like a lot of time and work. I don't think we should delegate a job like that to our subordinates."

"I agree it is a lot of work, but isn't it the real work we should be doing?" says Johnson. "I said we would develop an organization-wide strategy that can be used to implement the performance acceleration process. We all need to be deeply involved, and that is the reason I am focusing the executive team on this project. Along with your other responsibilities, each of you, as members of the executive team, will meet regularly to review the progress and to ensure that the activities are being properly prioritized."

A big moan goes up all around the table.

Tom Weston, vice president of research and development engineering, speaks up. "John, that doesn't include me, does it? I can see why manufacturing needs this program, but it really does not apply to my organization."

Johnson responds, "This is not just a program. It's a management style that will be used by every manager, and it's the performance standard that is acceptable for every area in the company. Your organization is no exception. But, I bet if you ask Rich [Rich Favor, vice president of manufacturing], he would tell you that you released the 915 unit without considering how it could be built and then tried to redesign it while it was in production. Just take time to look at your engineering change budget. That is your cost of doing the job over and over and over again. I understand that you are averaging seven changes per print during the product's life cycle. I think that's too many, don't you?"

"Yes, but—," replies Tom Weston.

"No buts about it," interrupts Johnson. "We all have our crosses to bear. Yours is engineering changes. Rich's is scrap and rework. Production control causes us to miss shipments because parts are not shipped from our suppliers on time, and then when they get here, quality assurance rejects them. The next thing I hear is that we are using the parts anyway because they are off-specs. When engineering tells us that it is all right to use parts that are not to the print requirements or just telling our suppliers that the prints are only guidelines, not a requirement. I was told by one of our suppliers' owner that often they hold parts they know are not to print, and ship them late rather than scrapping them out and making good parts because a high percentage of the time we will use them anyway."

"John," Mary Cross speaks up, "This looks like a QA problem. Why don't we let Bob Doright, our vice president of quality, take care of it and report back to us on what he is doing to solve it?"

"That's the attitude that got us into this present position," states Johnson. "QA has problems and I expect them to solve the ones they cause and report to us on product-related problems that we missed. But, what I'm talking about goes beyond the limits I have placed on quality assurance. These activities apply to everyone in the company. For every job we are doing, I want us to develop a system that prevents problems from occurring, not to detect and report them after they have occurred. If the rest of us are doing the right job of managing the business, we could almost do away with quality assurance. Each manager should ensure the quality of his or her area's output. We have too many quality assurance managers and too few quality managers."

Johnson paused for a moment. The room is quiet, so he continues, "I'm not just talking about the errors we are making; there's major room for improvement in the things we are doing right, the things that do not create errors, but take too much resources and time to accomplish. We've been making many good decisions, but not perfect decisions. When I look back at some of the decisions we've made, they were good decisions, but they could've been better. I realize that hindsight is 20/20, but we need to improve the way we look at the future and set in place plans that mitigate the risks that we're facing. I know that some of the things that we will be doing over the next few months may be new and strange to some of you. I'm going to ask you to accept these new concepts without questioning them. Likewise, I expect you not to reject them without giving them a fair chance to develop and mature in our environment. We are all good managers today; if we weren't, we wouldn't have the jobs we have. What I want from my management team is to go beyond good; I want us to be the very best we can be. Now I expect full cooperation and support from each of you in this important assignment. I will hold the first meeting of the Performance Acceleration Management Steering Committee a week from Thursday. It will be a two-day, offsite meeting. I will expect each of you to attend. This Steering Committee's activities are a top priority, so if any of you cannot attend, I would like you to call me beforehand and explain what has higher priorities. We have three driving factors in today's business. They are quality, cost, and schedule. In today's environment, balancing these three driving factors to maximize the value we provide to our customers is crucial for our survival."

Johnson directs the group's attention to two gentlemen entering the room. "Now I would like to introduce you to Jim and Frank, who are consultants that specialize in performance acceleration approaches. They will be scheduling two-hour meetings with each of you to collect data that will be used at our two-day, offsite meeting. They will also be scheduling focus groups with your middle- and first-line managers and employees to get their inputs on how we can accelerate our organization's performance. The inputs from each of you and from the focus groups will be kept anonymous even from me. So, feel very free to be open and candid in your inputs to these individuals. We have selected onsite consultants to collect this information, so that the inputs they receive will not be reflected negatively on any individual that volunteers criticism of the way the organization is performing today. I'm counting on each of you giving your full cooperation and candid inputs to these two gentlemen."

Johnson continues, "This meeting was scheduled for two hours. I'm planning on ending right on time. I will now ask Jim to give you a quick overview of the Performance Acceleration Management methodology that will last no more than one hour. That will leave 15 minutes for questions and answers. We have started Phase I of this methodology. If we like the results at the end of Phase II, we will move forward with its implementation."

Activity Two: Define Present Status and Improvement Opportunities

Definition

Key Performance Drivers (KPDs): Things within the organization that management can change that control or influence the organization's culture and the way the organization operates. (These are also called *controllable factors*).

During Phase I, Activity Two information related to the following tasks will be collected:

- Define the AS IS condition for each KPD
- Define the perceived level of conformance to the organization's value statements
- Define the organization's desire to improve
- Define opportunities for improvement
- Define the current culture of the organization
- Define current improvement projects that are underway

This information will be collected by conducting a series of one-on-one personal interviews with each of the executive team, focus groups with middle managers, first-line managers, and employees, plus surveys, and researching available documentation.

One-on-One Personal Interviews with Each Member of the Executive Team

Often this phase will start by the organization conducting a series of one-on-one, personal and confidential interviews with the members of the executive team. At this time, each executive will be asked to complete survey questionnaires and express his or her views on how the performance of the organization can be accelerated. Often a few middle managers, who have a wide view of the organization's performance, also will be included in this group. Typical middle managers that would be considered are the HR manager, Quality Assurance (QA) manager, Project Office manager, IT manager, and Customer Service manager. These private interviews are best conducted by individuals that are not members of the organization and have a reputation for keeping the confidentiality of the information they receive as it relates to its sources. Typical subjects that are discussed during these meetings include:

1. What are the roles and responsibilities and authorities of the individual being interviewed?
2. What roles and responsibilities will the individual being interviewed have in the PAM methodology?
3. How effective is the present organization's structure in bringing about improvements in organizational performance?
4. What are the opportunities for accelerated performance improvement?
5. What performance improvement programs are presently being used, have been used, and how effective are they?
6. What units within the organization have been the most successful in implementing performance improvement concepts?
7. What roadblocks do they see related to accelerating the organization's performance?
8. What is the level of stress related to the activities going on within the organization today?
9. What areas of the business are in need of the biggest improvements?
10. How would they describe the culture of the organization?
11. What are the KPDs for the organization?

Most organizations have only 8 to 12 KPDs. Typical KPDs that impact organizational performance include:

1. Measurement systems*
2. Training*
3. Management and leadership methods*
4. External customer partnership interface*
5. Supplier partnership*
6. Business processes*
7. Production processes*
8. Knowledge management*
9. Corporate interface to divisions
10. Employee partnerships
11. Research and development activities
12. After-sales service process
13. Information technology*

(The asterisk (*) denotes KPDs that are common to most organizations and have maturity grids developed for each of them.)

We suggest you conduct three types of surveys during these interviews:

1. Is/Should Be surveys
   These are surveys where a number of key questions are asked to be evaluated in three ways: on where it is today, where it should be five years in the future, and finally, from the viewpoint of the person taking the survey, how important is it to accomplish the higher performance level (priority). A typical question that would be asked is: "How much trust does the employee have in the management team?"
2. KPDs' Maturity Grids analysis
   For each KPD, a 12-level maturity grid has been developed and is used in the survey. The person completing the survey will be asked to read all 12 levels. Next, starting with the first and lowest level on the maturity grid, decide if he or she believes the total organization is in compliance with that level. If so, he or she is to go on and read the next level. This will continue until the individual taking the survey reads a statement describing a level that he or she does not believe the total organization has reached. At that point, the person taking the survey will circle the statement number by the last level on the maturity grid (the highest level) at which the organization is performing. This is defined as the "AS IS" performance level for the total organization as viewed by the individual taking the survey.

For those executives completing the survey, they should then repeat the process based on his/her evaluation of how the organization that reports to him/her is operating related to each of the specific KPDs. Figure 2.2 is a typical 12-level maturity grid survey analysis for management leadership and support.

3. Organization's Value Survey
   This survey is designed to evaluate the perceived level of compliance to each of the organization's value statements as viewed by management and the employees within the organization.

We suggest that you review each of these three surveys with the executives during these one-on-one meetings because their input can provide additional insight into their view of the organization. For example, in the "Is/Should Be survey," the executive may rate the employees' morale as "poor" at the present time and they would like to see a "very good" in the future, but rates a priority of making a change as very low. When the executive is asked why he/she gave it a low priority, his/her answer could be because he/she thought there was little chance of changing it. In the KPDs Maturity Grid analysis, it is always good to have a discussion with the executive on any point that he/she rated as very low or very high to understand the reason for the rating.

Focus Groups with Middle Managers, First-Line Managers, and Employees

Definition

Focus Group: A group of people, who have a common experience or interest, that are brought together where a discussion related to the item being analyzed takes place to define the group's opinion/suggestions related to the item being discussed.

In the past, most of the performance improvement programs have been driven by the executive team with little or no input from the rest of the organization. We understand that the executive team has the total responsibility for the organization's performance, but we also realize that the organization's operations and the needs of individual parts of the organization are viewed very differently at the different levels of management and from the employees' level. It is for this very reason that we feel input

Key Change Area

1. Management Support/Leadership

Scale

1. Managers give orders. Employees are responsible for following them exactly without question. Management gets credit for all successes; employees are blamed for failures.
2. Managers give orders. Employees are responsible for following them exactly, but are allowed to question them. Employees are blamed when they do not follow orders.
3. Managers are responsible for results; workers respond to the directives of management.
4. Managers recognize the need for change. Recognition/rewards begin to be a part of the motivation process. Managers start looking for and praising people who do the right things right.
5. Managers create a vision of the preferred future, which leads to group development of the "mission." An organization-wide plan for achieving the mission has been developed. All managers are trained in participative management techniques. Teams are formed to work on problem-solving and improvement opportunities.
6. A continuous improvement process is launched; team building and problem-solving training are provided to everyone. Managers recognize the need to be process-oriented. Progress has been made in building pride of accomplishment and self-esteem. Supervisors and managers are selected based primarily on their leadership ability.
7. Managers begin teaching, coaching, and working with their people on continuous improvement. Managers are treating quality and productivity as one. There are numerous examples of team building. An error-free performance standard is being used.
8. Management is working to change systems/processes, which their organizations have identified as barriers to achieving the organization's mission. "Management by Walking Around" is actively practiced. All employees are active members of a team. To improve, supervisors, and their teams, use employee surveys.
9. Managers tailor their organizations to facilitate continuous improvement. Quality and productivity performance levels and improvement "projects" are routinely reviewed with teams/individuals. All employees' output quality is measured and reported back to them.
10. Managers implement appropriate situational leadership concepts to stimulate groups and individuals in their groups in the implementation of a continuous improvement process. Managers are using statistical thinking. Teams are starting to set work standards. Promotions go to the people who prevent errors. A five-year plan that includes improvement activities is defined and understood by all.
11. Recognition and rewards clearly flow to those who are using a continuous improvement process. Firefighting is left to lower management and employees. Upper and middle management spend much of their time working with employees in their work areas, talking to donors, or doing long-range planning.
12. The culture includes the effective use of a continuous improvement process to continually improve quality, productivity, and employee morale. Workers are responsible for results; managers are responsive to their needs. Long-term quality goals are understood and supported by employees. Employees are setting their own time standards. The majority of management time is spent preventing errors.

FIGURE 2.2
Maturity grid for management leadership and support.

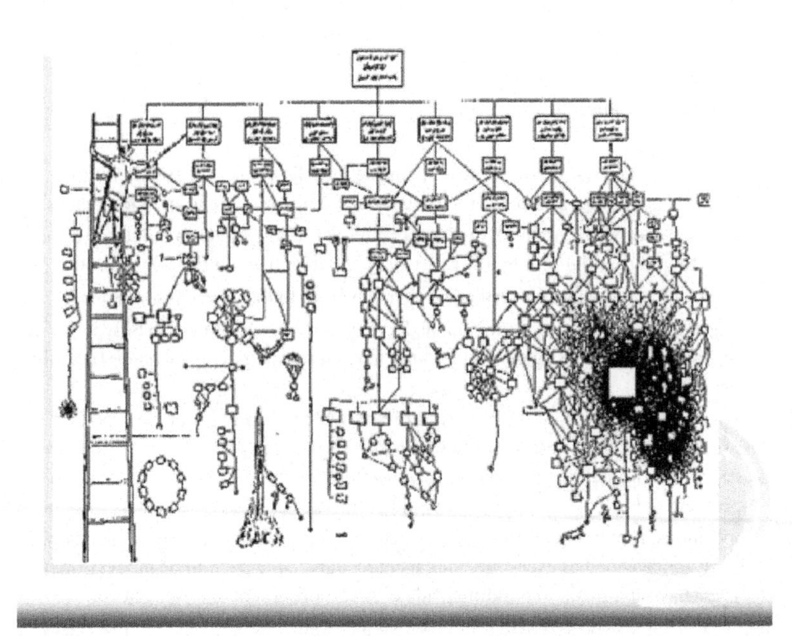

FIGURE 2.3
A typical organization chart as viewed by some employees.

related to performance acceleration is needed from all levels of the organization because frequently the view of the organization differs greatly based on the individual's vantage point. Figure 2.3 is a depiction of some employees' view of the organization chart in many organizations. And, Figure 2.4 provides a pictorial view of how some individuals at the bottom of the organization chart view the rest of the organization.

In the focus groups, a mixed population from the different functions provides a free exchange of information as viewed from different standpoints. Depending on the culture of the organization, this may be impossible and/or less productive than having a group from a single operating unit. One advantage of the focus groups is the ability to have an open discussion where the individuals involved exchange ideas related to an individual subject.

We like to conduct these focus groups in conjunction with the lunch break where we will bring in sandwiches or pizza to start the meeting off in a relaxed and friendly environment. The focus group is not intended to be a bitching session. The purpose is to have the participants express ways that would help them improve their quality and productivity. It is designed to help identify the roadblocks they are facing in their day-to-day activities. At the end of the focus group session, each participant will fill out an Is/Should Be survey and an organization's value survey.

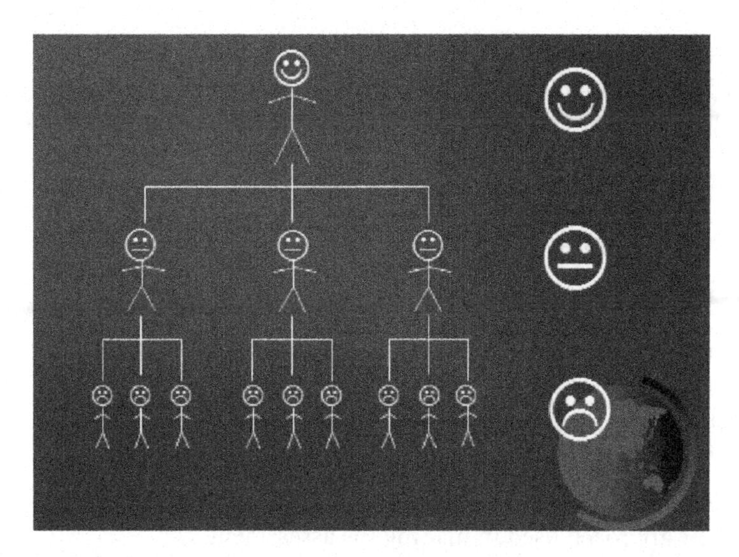

FIGURE 2.4
How individuals feel about their organization.

Researching Available Documentation

In most organizations, there is a great deal of previously published information that can serve as a rich source of performance acceleration opportunities. The individuals involved in Phase 1 need to search out and understand these documents. Some of the key documents that we find are valuable in identifying performance acceleration opportunities include:

- Employee opinion surveys and the corrective action plans related to these surveys
- Customer opinion surveys and the corrective action related to the surveys
- Strategic business plans
- Customer service desk records
- Supplier performance records and corrective action
- In-process yield reports and corrective action
- Customer complaints and corrective action
- Present approved projects' status reports

It is extremely important that the individuals conducting Phase 1 understand what improvement processes have been tried in the past and how effective or ineffective they were. It is equally important to understand the reasons that previous improvement efforts were not as successful as expected. No organization can afford duplicated activities or fail to learn from previous failures. It is costly to try something and fail, but it's unforgivable not to learn from the failure.

PHASE I FINAL REPORT

The information and understanding collected during Phase 1 should be documented in a final formal report. At a minimum, this report should include the following:

- Purpose of the assessment
- The approach used in making the assessment
- A list of the people involved in making the assessment
- List of the people that were interviewed during the assessment
- An overview of the results of the assessment
- Positive points revealed during the assessment
- List of the documents that were reviewed
- Survey results:
  - IS/Should Be survey
  - KPDs' maturity grids analysis
- List of improvement opportunities defined during the assessment
- Areas of concern
- Recommendations
- Typical comments made by the people being interviewed. Care should be taken to ensure that these comments can't be traced back to an individual.

SUMMARY

During this phase of the PAM project, an organizational structure was formed and trained. In addition, a comprehensive assessment was made to accomplish the following activities:

- Defining improvement opportunities
- Identifying roadblocks to implementing the PAM approach

- Identifying other improvement activities now underway
- Understanding why previous improvement activities were less productive than desired

The final report that documented the results of the assessment was prepared and will be used during Phase II as a basis to preparing five-year vision statements for each of the KPDs. It was recommended that this assessment be conducted by outside individuals in order to provide a higher level of confidentiality for the individuals being surveyed and providing more realistic uncensored data on which the assessment is based. This assessment will provide a baseline to measure the success of the PAM approach selected to be implemented over the next five years.

Many people are bewildered when the organization starts an improvement process and then become even more confused after it's implemented.

H. James Harrington

3

Phase II: Developing Vision Statements

> Vision statements that do not relate directly to what your employees are experiencing and/or doing have little or no impact upon their performance.
>
> **H. James Harrington**

OVERVIEW OF PHASE II

During this phase of the Performance Acceleration Management (PAM) approach, the organization will develop a separate vision statement for each of the Key Performance Drivers (KPDs). These vision statements will reflect the desired operational status of that KPD five years in the future. Phase II consists of eight activities:

- Activity One: Review and upgrade if necessary the organization's mission statement, values, and vision
- Activity Two: Review the assessment report prepared during Phase I
- Activity Three: Define the AS IS description for each KPD
- Activity Four: Develop preliminary KPD vision statements
- Activity Five: Conduct focus groups
- Activity Six: Conduct and analyze an Organizational Change Management survey
- Activity Seven: Prepare final KPDs' vision statements
- Activity Eight: Communicate the final KPDs' vision statements

Offsite Meeting of the Steering Committee

> The very essence of leadership is that you have to have a vision. It's got to be a vision you articulate clearly and forcefully on every occasion. You can't blow an uncertain trumpet.
>
> **Father Theodore Hesburgh**
> *Former president*
> *Notre Dame University*

We recommend that you start this phase of the PAM project by holding a two-day, offsite meeting with all the members of the Steering Committee in attendance. We like to plan the session so that the attendees stay overnight allowing them to socialize and get to know each other better. On occasions, we have made this a three-day, offsite meeting with the third day devoted to teambuilding including a golf match. As we will point out later in this chapter, the results from the assessment will give you a good indication if additional effort needs to be expended to build understanding and cooperation between the various functions within the organization. The following is a typical agenda for the offsite meeting.

- Comments by the highest-ranking officer attending the meeting
- Review of the meeting agenda
- Discussion related to the organization's mission, vision, and values
- Presentation and discussion related to the assessment report
- Prepare AS IS description for each of the KPDs
- Prepare a five-year vision statement for each of the KPDs
- Develop a plan to get input from the organization's stakeholders related to the preliminary vision statements

The Steering Committee offsite meeting starts with the highest ranking officer attending the meeting (president, CEO, COO, etc.) addressing the committee and pointing out how important it is for the organization to improve its performance at an accelerated rate over what it's been able to accomplish in the previous years. You should back up this statement by providing some examples of performance challenges that the organization is facing or that the organization will face in the coming years. The speaker also should stress that it is absolutely essential that the organization provide increased value to all of its stakeholders (customers, investors, suppliers, employees, and management). In addition, he or she should point out that the purpose of this offsite meeting is to provide the Steering Committee with an opportunity to define how the organization should operate in order to maximize its potential at an accelerated rate.

Activity One: Review and Upgrade, if Necessary, the Organization's Mission Statement, Values, and Vision

An organization's products, performance, and culture should be directed and driven by the organization's mission statement, value statements, and vision. Based on this belief, the Steering Committee should review and discuss how well these key organizational documents are serving the purpose for which they were created. If for some reason, they are out of date or not being complied with, this is an excellent time to discuss what needs to be changed in order to be in compliance with these key documents.

Definition

Mission Statement: The stated reason for the existence of the organization. It is usually prepared by the CEO and key members of the executive team. It is typically changed only when the organization decides to pursue a completely new market.

The mission statement is essentially a link to the organization with its vision of the future. Some organizations call this their "purpose statement" or the central reason why they are in business. A good mission statement will require leadership and be externally focused with the customer in mind and will serve as a motivating factor for individuals within the organization. It can be a "to be" or "to do" type statement. Both types of mission statements can be effective. Let the academicians or consultants argue over which one is correct; just use one that fits best with your culture and your team.

An example of a "to be" mission statement from Boeing is

> Our long-range mission is to be the number one aerospace company in the world, and among the premier industrial firms, as measured by quality, profitability, and growth.

An example of a "to do" mission statement from McDonald's is

> To satisfy the world's appetite for good food, well served, and at a price people can afford.

What is the difference between a vision and a mission for an organization? Many organizations go through an agonizing process of trying to determine the distinction based on the definitions offered by consultants and academics. Instead of trying to split the "definitional hairs," we have found that a useful perspective can be gained by utilizing the approach of linking your organization's internal effort with the external world in which you compete and serve customers. In this regard, having a compelling vision of the future of the market and the industry is essential. By *vision*, we mean a view of what the business will be like 10 to 20 years from now. It could be as simple as "an affordable, business, easy-to-use personal computer on everyone's desk" or "news available immediately from anywhere in the world." The winners tend to be able to express an energizing picture of the future in terms of market presence and customer benefits and have enough reality to it to make it aggressively believable. The losers tend to lack any vision and exist from day-to-day reacting to the market and the leads of other competitors.

Values can be defined as the deeply ingrained operating rules or guiding principles of an organization. Some may see these as specific culture-attribute statements that drive behavior. The winning organization sets out to create a specific culture and operating style to further define their strategic change and focus. They may be called basic beliefs, guiding principles, or operating rules. Call them what you will. The important thing is that they must be defined and the organization must live up to them because they serve as the "Stakeholders' Bill of Rights." There should be something in the organization's value statement that applies to the organization's relationship with each of its stakeholders. The following is a statement from IBM's Managers Manual defining IBM's principles:

> An organization, like an individual, must be built on a bedrock of sound beliefs if it is to survive and succeed. It must stand by these beliefs in conducting its business. Every manager must live by these beliefs in making decisions and taking action.

Now let's analyze IBM's principles to see how they apply to each of IBM's stakeholders:

- Respect for the individual (applies to IBM's employees).
- Service to our customer (applies to IBM's external customers).
- Excellence must be a way of life (applies to external customers and internal customers).
- Management must lead effectively (applies to the management team).
- Obligation to stockholders (applies to the stockholders who invested their hard-earned money in IBM).
- Fair deal for the suppliers (applies to the many suppliers that are essential for IBM to exist).
- IBM should be a good corporate citizen (applies to IBM's obligations and commitment to the community, nation, and the world).

It is obvious that IBM documented their commitment to their stakeholders as part of their basic operating principles. The Steering Committee should evaluate the organization's principles to be sure that they apply to all of the stakeholders of the organization. If they do not, there may be a need to modify them.

During the assessment activity in Phase I, we recommended that a survey be conducted that is designed to determine the perceived compliance to the organization's value statement. In this survey, the individuals are asked to evaluate each of the value statements to obtain their view of how well the organization is operating in compliance with the value statement. The survey is worded as follows: The following is a list of the organization's value statements. From your perspective, do you agree or disagree that the organization is operating in keeping with these value statements?

They are given the following choices related to each value statement:

1. Strongly agree
2. Agree
3. Disagree
4. Strongly disagree
5. Have no opinion

The survey also asks the question: Were you aware of these value statements prior to taking the survey? Yes or No.

If the survey is conducted, it will provide a great deal of value-added insight during the discussions related to the present value statements. Based on past experience, the perceived degree of compliance to these value statements decrease as you go down in the organizational structure. The Steering Committee should be concerned if any group has 10% of the respondents answering negatively to any of the questions.

Activity Two: Review the Assessment Report Prepared during Phase I

Although the assessment report should have highlighted a number of positive points related to the organization, the primary purpose of this discussion should focus on identifying improvement opportunities. Therefore, the surveys that were conducted during the assessment will provide valuable insight to the Steering Committee's discussions.

The "Is/Should Be survey" provides the means of prioritizing the improvement needs related to the questions that were asked. Typically, the five questions that have the highest improvement priority for the executives, middle management, and employees are highlighted. Figure 3.1 is an actual example that identifies the five areas rated by the executive team as requiring the most improvement. It also identifies how different levels within the organization rated the same questions.

In this case, there were a total of 22 questions in the "Is/Should Be survey." You will note that the totals for middle management and employees exceeded 40, which mean that there is a problem because the executives do not understand middle management and employees' concerns. Using a specific example, the no. 4 priority for the executive team was: "How well do departments cooperate?" This was next to the lowest priority for middle manager and a very low priority for the employees.

Figure 3.2 shows the five areas rated by the employees in the same organization as requiring the most improvement. It also identifies how different levels within the organization rated the same questions.

You will note in this case that the executive total of 43 exceeds the 40, indicating that this is an area of concern. For example, the no. 1 priority for the employees was: "How much of managers' and employees' salary and rewards are based upon the quality of their work?" For the executives, this particular question was given a low priority rating, whereas the employees felt it was the no. 1 issue. The middle management total of 30 exceeds the 25 number, indicating that additional

The following identifies the five (5) areas rated by Executives as requiring the MOST improvement. It also identifies how the other demographic groups rated the same question.

Ratings

Executive	Mid-Mgmt.	Employee	Question
1	3	11	To what degree does your company focus on preventing errors? (1)
2	2	4	How good is employee morale? (7)
3	15	7	How well do all employees understand company long-range (5-year minimum) plans? (12)
4	21	16	How well do departments cooperate? (5)
5	1	3	How much trust and confidence do employees have in the management team?(6)
15	42	41	**Totals**

If very close agreement exists between the personnel groups, all totals should equal or be only slightly higher than 15. If the total of either of the other groups exceeds 25, it indicates that additional work is needed to gain better understanding between the two populations. Any total that exceeds 40 is an area of concern.

FIGURE 3.1
The executive team's five areas that require the most improvement compared to middle- and line-managers' and employees' priorities.

work needs to be expended to gain better understanding between the two populations.

This survey should provide the Steering Committee with an understanding of the need to service the improvement needs of all levels of the organization. The transformation should service the needs and concerns of all levels within the organization if they are going to sell the performance acceleration initiative.

In analyzing the KPD Maturity Grid Analysis, it is helpful to construct a graph that shows how the individual executives responded to the KPDs' maturity grid statements. Figure 3.3 shows how the executives at one of our clients responded to the maturity grid that is related to management support and leadership.

You will note that in Figure 3.3 the spread of the main population is four statements. In a cohesive executive team, the spread should be no more than three statements. Although a spread of four statements is an area of concern, when you look at the total population, the spread is a total of

The following identifies the five (5) areas rated by Employees as requiring the MOST improvement. It also identifies how the other demographic groups rated the same question.

Ratings

Employee	Executive	Mid-Mgmt	Question
1	17	5	How much of the managers' and employees' salaries and rewards are based on the quality of their work? (9)
2	7	8	How well are employees trained to do their job before they start providing output to their customers? (16)
3	5	1	How much trust and confidence do employees have in the management team? (6)
4	2	2	How good is employee morale? (7)
5	12	14	How well does management keep the employees informed? (19)
15	43	30	**Totals**

If very close agreement exists between the personnel groups, all totals should equal or be only slightly higher than 15. If the total of either of the other groups exceeds 25, it indicates that additional work is needed to gain better understanding between the two populations. Any total that exceeds 40 is an area of concern.

FIGURE 3.2
The employees' five areas that required the most improvement compared to the executives' and middle management priorities.

seven statements. Even worse, there is a second small population grouped around statements 7 and 8. This is an indication that the management style and activities are developing two separate and different cultures in various parts of the organization. This can present a major problem in the way the organization is functioning.

Figure 3.4 is a plot of how the executive team views the management support/leadership of the total organization. It is often interesting to plot as well how the individual executive feels about the maturity of the management support/leadership within his or her organization. Often the individual's evaluation of his/her organization is rated at a much higher level on the maturity grid than his/her evaluation of the total organization. This can lead to some very informative discussions related to what the individual executive's organization is doing to cause it to be rated higher on the maturity grid and how this can be applied to other parts of the organization.

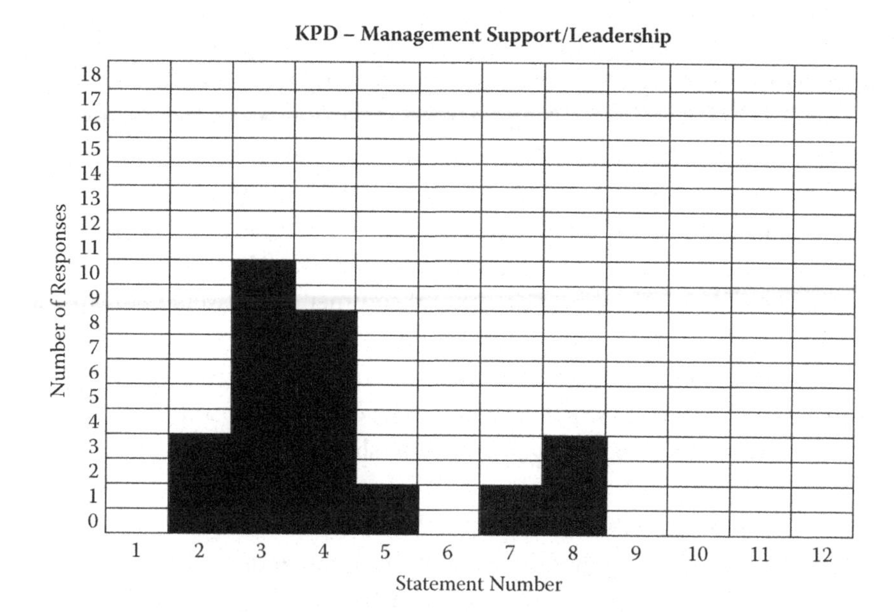

FIGURE 3.3
Maturity grid analysis for management support and leadership for the total organization.

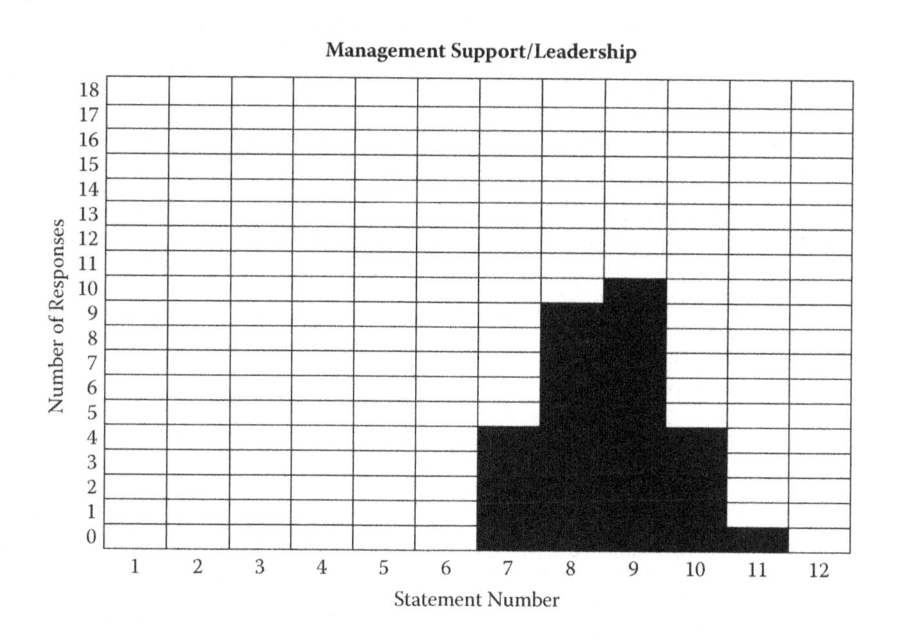

FIGURE 3.4
Plot of how the executive team views the management support/leadership in their area of responsibility.

Because the assessment team had confidential discussions with the key executives and conducted focus groups with middle managers, first-level managers, and employees (plus they had access to all of the key performance documents (employee opinion surveys, customer complaints, production yield reports, etc.)), they are in a good position to make a number of very meaningful observations and recommendations. This part of the assessment survey provides excellent input into the PAM database.

Also included in the report are typical comments that were made by the people interviewed. These comments provide additional insight into the improvement opportunities that the organization has. The following are some actual comments that we have received when we were conducting assessments related to management/leadership.

- Communications is a major problem.
- The family feeling has gone out of the organization.
- People are afraid of what will happen next.
- There is another issue that will drop pretty soon. We are all waiting to hear it.
- We get told anything they want to tell us. They don't mean it.
- They cut the pay 25% because future sales look bad, but management bonus went up. Does that make sense?
- We have lost the loyalty of our employees.
- If I was asked to work on the weekend, I would figure out some reason why I couldn't, just to get back at them.
- I need respect. I don't get it.
- There is a lot of turmoil in the organization.
- It's all a numbers game.
- They're not honest with us.
- Very poor leadership around here.
- They asked us what we think, but never take action on it.
- Management needs to get out and be seen.
- The company speaks with forked tongue.
- There is no cooperation between management.
- We need to trust our employees if we want their trust.

There is no doubt about it, when an assessment is conducted in a professional manner and the data is analyzed thoroughly, the Steering Committee will be provided with very important data that can be used to develop the organization's combined PAM plan. The real key to a successful and meaningful assessment is the ability of the assessors to do the interviewing and gain their confidence. This information must be kept confidential. Unfortunately, most engineers, and other support personnel, do not have adequate training in how to conduct an interview. They lack the skill to probe down to get to the meaning behind the various comments. They often ask questions that can lead to abrupt answers, such as "yes," "no," "I agree," or "don't know." These interviewers need to ask questions that require the interviewee to explain his or her reasoning. For example, asking the question: "What do you like most about working here?" provides an opportunity for more feedback than simply asking: "Do you like working here?"

Activity Three: Define the AS IS Description for Each KPD

The Steering Committee will now agree on written descriptions that define the present status (AS IS status) for each of the selected KPDs. The written description should be no longer than three sentences and two is even better. We do not recommend that a great deal of time be devoted to establishing an exact definition of the AS IS status. Often organizations that use the KPD Maturity Grid Assessment can use the definition defined by the average value of the input from the executives with no change or with just a slight modification. For KPDs, where there is a great difference in the executives' view of the AS IS status, we recommend that the Steering Committee be divided up into small working groups of three to five people that will address individual KPDs using the following agenda:

- Select team leader.
- Select a scribe.
- Brainstorm for key words and phrases that describe today's status.
- Have each team member write a statement that describes today's status, using selected words and phrases.
- Have each member read his/her statements.
- Have each person select his or her first and second choice of the statements prepared by the group members. Give two points for every first choice and one point for every second choice. Add up the scores for each statement. Select the two statements with the highest scores.
- Record the two selected statements on flip charts.
- Using the two selected statements and the total list of the selected word phrases, write a new statement that the group will fully

support. Remember, the statement represents the personality of the total organization, not an individual department or function. If you cannot get consensus, record both options.

- Select someone to present the results to the Steering Committee.
- Review the proposed statement with the Steering Committee and modify it until there is common agreement.

Activity Four: Develop Preliminary KPD Vision Statements

Now is the time to think about the future and how the organization and its stakeholders would most benefit from changes to the KPDs. The objective of this part of the offsite meeting is to look out five years in the future and prepare a verbal description of how each of the KPDs should be operating if the organization had significantly accelerated its performance improvement activities. The statements will be the desired future state for each of the KPDs. (It is also referred to sometimes as the *desired future state.*) These two- to three-sentence statements related to each of the KPDs are very important and should be prepared with a great deal of thought and discussion as they will drive the future culture of the organization. The problem with many organizations today is they haven't taken the time to define how they would like to modify the culture within their organization; instead they blindly accept today's "fad" to bring about performance improvement without considering its impact upon tomorrow's culture. It's a lot like driving down the road at 100 miles an hour with mud all over your front window and not being able to see whether the road is turning to the left or to the right (Figure 3.5).

To develop these KPD vision statements, the Steering Committee should be divided into small groups to develop a statement for each of the KPDs. It is very important that these small groups keep in mind the organization's mission, vision, and value statements as they prepare the individual KPD's vision statements. These statements must be in line with the organization's values and they should be directed at escalating the performance of the organization from the present AS IS status to the desired future state.

The mission of these small groups is to define how the organization's personality/culture should develop over the next five years related to the assigned KPD. The definition should be no longer than three sentences (and two is even better). The five-year vision statement should clearly define the desired way the organization should be operating five years from now. A good mission statement should:

FIGURE 3.5
Having a clear vision of where you are going.

- Be phrased in the present tense. (Example: "We are effectively using the participative management methods in all areas," not "We should be effectively using participative methods in all areas.")
- State what we are doing, not how we are doing it. (Example: "We are using statistical process control in all appropriate areas," not "We are using P charts in the plating departments."
- Show a state of change that corrects today's problems as defined in the employee opinion surveys, other key documents, and the assessment report.

A typical agenda for the small group meetings would be to:

- Select team leader.
- Select a scribe.
- Use the KPD Maturity Grid, if one is available, that best fits the desired future state.
- Add to this list by brainstorming for other words and phrases that describe the desired future state.
- Have each team member write a statement that describes the desired future state, using selected words and phrases.
- Record each statement on a flipchart.
- Have each member read and discuss his/her statement.

- Using the Delphi technique, select the two best statements. This may mean that you may have to cycle through the ratings more than one time.
- Using the two selected statements and the total list of words and phrases, write a new statement that the group will fully support. Remember, the statement represents the personality/culture of the total organization, not the individual department. If you cannot get consensus, record both options.
- Select someone to present the results to the Steering Committee.
- Present the vision statement to the Steering Committee and discuss it with the committee. During these discussions, it may be necessary to modify the statements until a consensus is reached.

It is very important that the Steering Committee gain a common understanding of the meaning and intent of each word that makes up each vision statement. Words like *empowerment* need to be discussed so there is common agreement on exactly what it means. Often the meaning of individual words needs to be recorded in writing so that their purpose and direction is commonly understood by the Steering Committee and can be communicated concisely and consistently to the rest of the organization. These vision statements are labeled as *preliminary* vision statements because they usually change significantly when inputs from the other stakeholders of the organization are included in the vision statements. A typical preliminary vision statement for business process could be

> Our critical processes provide a competitive advantage that employees and customers recognize as world class. Continuous improvement using statistical data is a way of life. Processes are defined with assigned owners, evaluated against established measures, and operate in the real world.

Activity Five: Conduct Focus Groups

> When people are involved and influence decisions, they make things happen. When they are not, they make sure things will not happen.
>
> **H. James Harrington**

In Activity Four, we described how the Steering Committee, which is made up mostly of executives, develops the preliminary vision statements for each KPD. These statements reflect the way management interprets the data they have and the way they picture the transformation of the organization's KPDs. However, management is only a small part of the people who are affected by these vision statements. There are three more stakeholders who also need to influence these vision statements. They include:

1. The customers
2. The employees
3. The suppliers

Each of the executives should take the preliminary vision statements back to their function and hold a series of focus group meetings with their direct reports, first-line managers, and employees. Each focus group should review all of the vision statements to determine the following:

1. Is this the type of environment you want to work in?
2. Is this different from today's environment?
3. Do you understand the vision statements and what each word means?
4. How could it be improved?
5. Do you think it is attainable?
6. What would keep us from achieving it?
7. What can you do to help improve the visions?

Often these sessions are kicked off by the person, who did the assessment, reviewing the findings. This helps everyone obtain a better picture of the AS IS condition. Flipcharts should be used to record all comments—negative or positive. This is an effective way to document the discussions and come to common agreement on key issues. Sometimes things look different in print.

At the end of the focus group meeting, the executive will thank those in the group for their input and tell them that these will be reviewed with the Steering Committee. The executive must point out that although all input will be discussed and considered, he or she cannot guarantee that they will be incorporated into the final formal vision statement.

Procurement should ask the major suppliers to attend a focus group meeting to review all of the vision statements; however, most of the supplier focus group's time should be dedicated to the supplier partnership vision statement. Marketing and/or sales should do the same thing for major customers, with particular emphasis focused on obtaining their customers' input related to the customer partnership vision statement. With both the supplier and customers, it is better to review the vision statements with too many, rather than too few.

This is one of the most exciting parts of the PAM project. In most organizations, it is the first time that management has ever asked the employees what type of environment in which they would like to spend their lives. Even if the executive team only talks to 2 or 3% of the employees, the word spreads like wildfire throughout the organization. It is too bad that, because most organizations have a major problem with employee trust, the typical comment is: "Looks great, but you have to prove it to me."

> Early in the history of the company, while thinking about how a company like this should be managed, it kept getting back to one concept: If we could simply get everyone to understand what we are trying to do, then—starting with people who want to work and providing them with the right conditions and resources to do it—we could turn everyone loose and they would move in a common direction.
>
> **Dave Packard**
> *Co-founder*
> *Hewlett-Packard*

Activity Six: Conduct and Analyze an Organizational Change Management Survey

How an organization reacts to a new change initiative is largely based on what experiences the people within the organization have had related to past change initiatives. Organizations that have been successful in bringing about change in initiating new projects will have far less resistance to bringing about additional changes in the organization's culture than those organizations that have failed in these efforts in the past. What goes on within the organization conditions its employees in their reaction to future similar activities. It, therefore, is essential to conduct an Organizational Change Management (OCM) survey to define which current behaviors will support change and which will resist change.

The inability to fully implement decisions that affect large numbers of people throughout an organization is a critical issue facing many organizations today. Major decisions with strategic implications, such as reorganizations, new policies, new procedures, or the introduction of new technology, will fail when an organization lacks the capability to translate senior-level directives into tangible results. One of the key factors that affect an organization's ability to implement present or future changes is its past implementation success. Previous experience is a potential predictor of the future. Past implementation difficulties are likely to be repeated when engaging in a new project. Therefore, an assessment of previous implementation experience is critical to planning for future changes. An OCM survey would normally be conducted when the following conditions exist:

- While organization change is being considered or during initial planning
- Before the change has been announced
- Any time after an announcement has been made
- After project implementation is complete

Why:

- To provide early warning of potential problems and possible implementation failures
- To determine the organization's predisposition toward change
- To analyze any barriers that may arise during the implementation process
- To identify new barriers that may have developed after implementation

An ideal time to conduct an OCM survey and analyze the results is when the focus groups are being conducted. At a very minimum, all of the members on the Steering Committee should complete the survey. In addition, it is good practice to survey the individuals that make up the focus groups. This will provide you with two to three levels of opinions related to the problems the organization will have in bringing about a major change in the organization's personality/culture.

By carefully analyzing the results of the OCM survey, the organization will be provided with a list of things that are considered enablers to bring about a successful implementation of the transformation. There is another list of items that are high risk or barriers that will have a negative impact on the successful implementation of the project. These are the items that, in mitigation plans, need to be developed to minimize their negative impact upon the project.

The results of the OCM survey should be presented to the Steering Committee when it meets to prepare the final formal vision statements.

These results also should be considered as the individual KPD transformation plans and the combined PAM plans are being prepared.

> Employees do not expect management to implement every suggestion they make, but they do expect management to listen to them and consider the ideas they have on items that impact them.
>
> **H. James Harrington**

Activity Seven: Prepare the Final KPD Vision Statements

When the results of all the focus groups are available, a second offsite meeting is scheduled for the Steering Committee. During this meeting, the agenda will include the following:

1. Preparing the final KPDs' vision statements.
2. Setting accelerated performance improvement goals.
3. Defining desired behavior and habit patterns.
4. Assign a team leader to assemble the subcommittee for each KPD to develop a three-year individual transformation plan.

In preparing the final KPDs' vision statements, the executives not only represent themselves, but they represent their employees by providing feedback to the Steering Committee on the results of the focus group meetings. We have participated in many of these sessions and without exception there have been major changes made in the preliminary vision statements that were prepared by the Steering Committee when the inputs from the focus groups were considered. These are very important changes because they provide visible evidence that the executive team is listening to and taking action regarding their employees' concerns and inputs. Even though an individual employee did not have his or her suggestion included, he or she was able to see that the executives changed their vision statements after they talked with the employees. The following is an example of how the preliminary vision statement was altered during one of our client's offsite sessions based on the inputs the Steering Committee received as a result of the focus group meetings.

> **Preliminary Vision**—Our critical processes provide a competitive advantage that employees and customers recognize as world-class. Continuous

improvement using statistical data is a way of life. Processes are defined with assigned owners, evaluated against established measures, and operate in the real world.

Final Vision—Our processes are recognized as world-class—innovative, streamlined, effective, and adaptable—by both internal and external customers. We define our processes, empower the process owners, and apply effective measurements.

Activity Eight: Communicate Final KPD Vision Statements

The full potential of our employees emerge when meaningful, common visions are created and agreed to.

H. James Harrington

Management's job is to promote these agreed-to vision statements. This does not mean to only talk about them or to support them, it means to live them, to sell them, to be enthusiastic about them. Confusion reigns supreme when managers talk and write one message, but act and live another. If management cannot live and behave in keeping with the vision, for heaven's sake, don't talk about or document them. It is for this very reason that the Steering Committee should set higher priorities on completing Phase III: Defining Desired Behavioral Patterns and Performance Goals and complete the planning for Phase IV: Developing Individual KPD Transformation Plans very quickly.

Once the final set of KPDs' vision statements have been agreed to, a communication plan needs to be prepared. The organization also must define how this information will be disseminated to all of its employees within the organization and its key suppliers and customers outside of the organization. For disseminating the information internally, we find it is effective to show both the preliminary vision statements and the revised final vision statements. This reinforces the fact that the executive team listened to the employees and agreed to modify the vision statements to reflect their inputs.

Too many organizations have 20/20 vision when they are looking backwards, but are blind as they go forward.

H. James Harrington

SUMMARY

During this phase, the Steering Committee has developed a set of vision statements using input from their employees, suppliers, and key customers that it will use to define the future desirable internal operations and culture of the organization. This is an activity that has to be taken in a very serious and thoughtful manner because it will be used to drive the organization's future operations and will have a major impact on all of its stakeholders. These vision statements will drive the behavioral patterns of the executives and employees to bring about a major change in behavioral patterns and culture. If these visions are properly defined, they will bring about a major acceleration in the organization's performance and have a very positive impact on all of the organization's stakeholders.

These vision statements turn out to be very important commitments that the organization is making to its employees, investors, suppliers, management, and its external customers, as well as all of its potential customers and investors. This is an exciting time within organizations because it is for most organizations the first time that management has ever asked and invited their employees to participate in defining the environment in which they will be working. This sets high levels of excitement and expectation throughout the organization that can easily be turned into high levels of dissatisfaction if there isn't immediate and visible activities implemented to bring about the desired transformations.

> The worst thing that an organization can do is to make a commitment and not live up to it.
>
> **H. James Harrington**

4

Phase III: Defining Desired Behavioral Patterns and Performance Goals

> Everyone must change the way they behave before they can change the way they perform.
>
> **H. James Harrington**

INTRODUCTION

In order to accelerate the performance of the organization, first of all, management and employee behavioral patterns need to change, and then these changes will be reflected in accelerated performance results. Based on this reality, there are two very distinct and different results that need to be targeted for change:

1. The behavioral patterns of management and the employees need to undergo a significant change and that change has to be sustained long enough so that it becomes a habit (or to put it another way, it becomes "the normal way it's done here").
2. If the desired behavioral patterns are established, then measurable results will be obtained and can be sustained.

We, therefore, have separated Phase III into two distinct and separate parts: (1) Defining the Desired Behavioral Patterns, and (2) Defining Performance Goals. There are a total of eight activities in Phase III broken down as follows:

1. Defining Desired Behavioral Patterns:
   a. Activity One: Define the desired behaviors/habits related to the vision statements
   b. Activity Two: Define the desired and undesired behaviors/habits related to the activities as performed within the organization
   c. Activity Three: Define how to measure desired behavioral patterns
2. Defining Performance Goals:
   a. Activity Four: Define key performance measurements
   b. Activity Five: Define present performance levels of the key performance measurements
   c. Activity Six: Define competitors' present performance level for the key performance measurements
   d. Activity Seven: Project competitors' performance level for the key performance measurements
   e. Activity Eight: Finalize the key performance goals for the next five years

> If you focus on meaningful long-range goals, you can overcome short-range obstacles.
>
> **H. James Harrington**

DEFINING DESIRED BEHAVIORAL PATTERNS

> Life asks us to make measurable progress in reasonable time. That's why they make those fourth-grade chairs so small.[1]
>
> **Jim Rohn**
>
> *Excerpts from The Treasury of Quotes*

Once the desired behavioral patterns have been defined, a list of these behaviors should be prepared and prominently displayed within the organization. It is important to note that changes in behavioral patterns can be observed prior to improvements in key performance measurements, such as customer satisfaction and reduced cycle time. There are three activities involved in defining the desired behavioral patterns:

- Activity One: Defining the desired behaviors/habits related to the vision statements
- Activity Two: Defining the desired and undesired behaviors/habits related to the activities as performed within the organization
- Activity Three: Defining how to measure desired behavioral patterns

Immediately after firming up the final set of Key Performance Drivers (KPDs) vision statements and while the discussions related to each of them is fresh in the minds of the Steering Committee, the members should define what are the desired results from the Performance Acceleration Management (PAM) process. (This is normally done as the second part of the agenda for the second meeting of the Steering Committee.) The start of the personality/culture change in an organization is developing a set of vision statements. If they are worthwhile and they are embraced by management and the employees alike, then the individual's feelings and thought patterns will begin to change. If the organization and the individuals involved are rewarded personally and socially as these new culture changes are embraced, over time they will transform into normal behavior and habit patterns. For example, if part of your Management Support/Leadership vision statement was to "empower employees at all levels," this part of the vision could be reflected first in your employees as they begin to feel that they don't have to get management approval to take action on unplanned-for events. This will allow the employees to begin to gain confidence that they can make the right decision in many cases without management's help. With time and continuing management support, they can begin to feel confident that they will not be reprimanded because they made a decision, and, thus, their behavioral patterns will change. More and more, they will take the needed action, often telling management after the fact about the problem and how it was handled. They will start to come to management, explaining how they are going to correct the problem instead of asking management how to solve it. Positive reinforced behaviors and action become habits. At this point, these specific patterns become a normal way of doing business and the sentiment will turn into: "It's just the way we do things around here; it's nothing special."

> Behaviors drive results. When changed behaviors become habits, the culture is changed.
>
> **H. James Harrington**

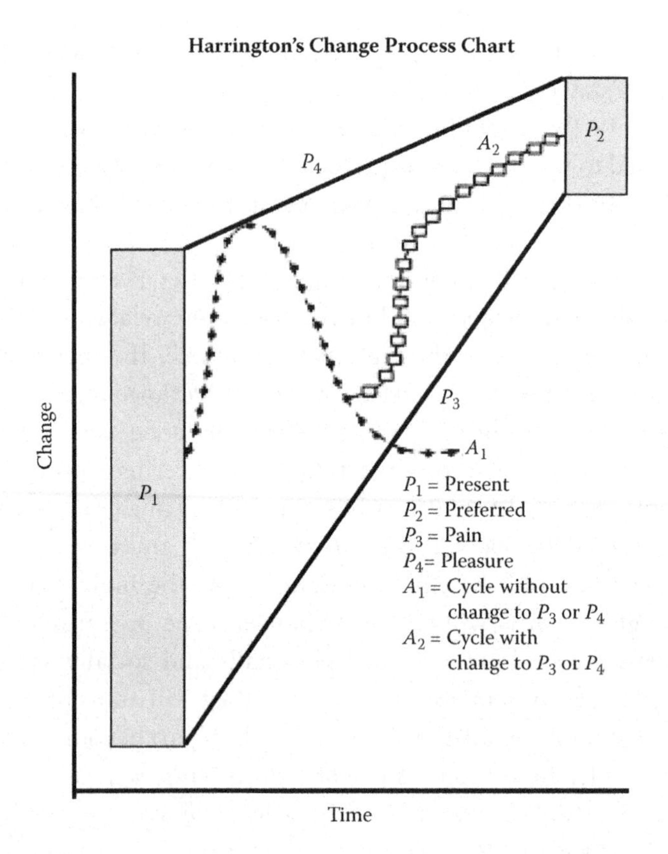

FIGURE 4.1
Change process chart.

In some organizations, this change in behavioral patterns is so drastic that behaviors that were previously rewarded become behaviors that are now considered undesirable (Figure 4.1).

In Figure 4.1, P_1 stands for the present behavioral pattern within the organization. P_2 represents the desired behavioral pattern. You will note that P_2 has a much higher performance level and has less variation than P_1 throughout the organization. P_3 stands for pain and it's the point where the organization needs to take action to turn around undesirable behaviors and/or activities. This is usually some type of discipline. P_4 stands for pleasure. It is this point in a behavioral pattern where the individual is performing and/or behaving above expectations and is rewarded for behaving in the unusually acceptable manner. A_1 is a typical cycle that the organization will go through when initial action is taken to change the present behavioral pattern without a change in the pleasure or pain points.

You will note that there is an immediate positive change in the average behavioral pattern, but over time, as the newness of the change wears off, the behavior pattern slips back to where it was originally. A_2 is the average behavioral pattern of the organization resulting from the pleasure point to be recognized for superior performance that continuously becomes more stringent. A_2's positive progress also is driven because P_3 (the pain point where action is taken to stop unacceptable behavior) is continuously set to require a higher level of conformance to the desired behavior. In the example shown in Figure 4.1, the pain point eventually reaches the point where it requires that the organization perform at a higher level than its average behavioral pattern in P_1 (the present status). It is for this very reason that it is absolutely essential to set a new reward and recognition system in place, as well as establish a disciplinary system, in order to sustain the long-term changes that are required in the behavioral patterns.

Activity One: Define the Desired Behaviors/ Habits Related to the Vision Statements

> I used to say, 'I sure hope things will change.' Then I learned that the only way things are going to change for me is when I change.[1]
>
> **Jim Rohn**
> *Excerpts from The Treasury of Quotes*

During the first part of Phase III, the Steering Committee is responsible for preparing a list of desired behavioral patterns and determining how these behavioral patterns are demonstrated in the day-to-day activities within the organization. At a later phase in the PAM project, the organization's reward and recognition system should be changed to recognize the desired behavioral patterns, so that these patterns can be reinforced.

To start to define desired behavioral patterns, the Steering Committee should review each of the KPDs' vision statements and identify words/ phrases that drive the behavioral patterns within the organization. Typical examples include:

- Empowered workforce
- Adaptability
- Bureaucracy-free
- Technology-driven

- Customer-driven
- Process focus
- Streamline operations
- Fast response
- Quality first
- Minimum waste
- Continuous career enhancement
- Job rotation
- Free knowledge exchange
- Foster self esteem
- Trust and respect at all levels
- Open communications
- Participative management
- Technology minimizes routine repetitive activities
- High levels of creativity
- Leaders at all levels
- Wild ideas are encouraged and discussed
- Managers are considered angel advocates, not devil advocates. Their mission in life is to help develop the employees' idea rather than tell them why it will not work.

Using one of the most popular phrases related to the Management Support/Leadership vision statement—empower the workforce—the following are some of the behaviors and patterns that would be observed in an empowered workforce:

- First-level employees define their work process and time schedules.
- There is less second-guessing.
- Decisions are made more quickly and at lower levels.
- Management defines results expected, not how to get them.
- Unsolicited recommendations and suggestions are offered by the first-level employees.
- Self-managed work teams are used effectively.

Excellence is an art won by training and habituation. ... We are what we repeatedly do. Excellence, then, is not an act but a habit.

Aristotle

Activity Two: Define the Desired and Undesired Behaviors/Habits Related to the Activities as Performed within the Organization

In addition to selecting keywords and phrases from the vision statements, the Steering Committee should define repetitive activities that go on within the organization and identify desirable and undesirable behavioral patterns related to these repetitive activities. The following is a list of some of the more common undesirable behavioral patterns that exist in many organizations today:

- People showing up for meetings late.
- Meetings not ending on schedule.
- People coming to work late.
- People taking long lunch hours.
- Unusually high absentee rates.
- Excessively long breaks.
- Meetings with no agendas.
- Use of the Internet during company hours for personal activities.
- People who arrive at work on time, but spend extensive amounts of time before they start doing productive work. (For example, conducting personal meetings in the coffee break room discussing last night's television programs.)
- Not meeting commitments.
- Discrimination and harassment.
- Cluttered work areas.
- Hoarding of information.
- Lack of documentation related to the meetings that were held.

The Steering Committee should make a list of these desirable and undesirable behavioral patterns and identify the ones that they want to address over the next five years. For each of the identified unacceptable behavioral patterns, the Steering Committee needs to define what the desired behavior should be. In many cases, standards need to be established related to the behavioral pattern. (For example, an organization might set the following standard: It is unacceptable for anyone to show up more than five minutes late for a meeting unless the individual had contacted the person scheduling the meeting in advance notifying him/her that he/she will be late.)

It is the responsibility of every member of the Steering Committee to be a role model by living up to the behavioral standards they are defining for the rest of the organization. One of the most effective ways to bring about organizational change and accelerated performance improvement within an organization is for the executive team to start measuring themselves in keeping with the behavioral patterns that they expect the rest of the organization to maintain. We have implemented this concept in a number of organizations and the executive team has been amazed to find out that they were making between 50 to 80 behavioral errors per week at the start of the program. As they focused on improving their behavioral patterns, they were able to reduce this down to less than 10 per week. The result of this change in the executive team's behavioral patterns was startling, as it was reflected all the way throughout the organization and a new and much more positive operating personality penetrated the total organization.

Activity Three: Define How to Measure Desired Behavioral Patterns

The Steering Committee is now in a position to select key behaviors and establish ways of measuring how they need to change within the organization. For example:

- To measure how often unsolicited recommendations and suggestions are turned in, the organization can review the number of performance improvement ideas and suggestions that are turned in per each eligible employee.
- To measure the number of meetings starting on schedule, the person preparing the minutes of the meeting should record the scheduled start time of the meeting and the actual start time.
- To measure meeting commitments, the original committed target date should be recorded along with the actual completed date. Often the records only show that a target was completed, but it does not show the slippage in the original completion target date.

Using this approach to reinforce the desired behavioral patterns, the Steering Committee should develop an extensive list of behavior and habit pattern measurements, most of which are not measured in today's balanced scorecard. This list of measurements should be reviewed and many of them included when the Steering Committee creates its measurement plan.

DEFINING PERFORMANCE GOALS

> Five-year objectives in a highly dynamic environment are more a vision than a highly detailed set of objectives. It consists of one or two very clever differentiating concepts. A vision that stands the test of time and contains very good differentiators, combined with excellent year-to-year execution, is the best formula for business success.[2]
>
> **Dick Hackborn**
> *Peripherals Group vice president and general manager*
> *Hewlett-Packard*

During this second part of Phase III, the Steering Committee will establish yearly, high-level performance goals that are designed to make the organization more competitive than its competition. These goals are designed to accelerate the organization's performance over the next five years. They are used to prioritize the accelerated performance initiative so that they focus on the key performance measurements of the total organization. This part of Phase III consists of five activities:

- Activity Four: Defining key performance measurements
- Activity Five: Defining present performance levels of the key performance measurements
- Activity Six: Defining competitors' present performance level for the key performance measurements
- Activity Seven: Project competitors' performance level for the key performance measurements
- Activity Eight: Finalizing the key performance goals for the next three to five years

Activity Four: Define Key Performance Measurements

This set of key performance measurements is not meant to replace the balanced scorecard that is used to measure the total organization's performance. The purpose of these key measurements is to focus on setting accelerated improvement goals for the organization as they relate to the key stakeholders. It is less focused on financial results and more focused on changes in behavior and attitudes. It is primarily used to help define when individual improvement activities need to be introduced into the organization over the coming five years thereby setting the priorities for each of the performance acceleration improvement activities. Based on our experience, it is essential that any improvement initiative results in financial success early in its implementation. This is necessary because two of the key stakeholders' (management and investors) primary focus is on quarterly financial returns. Changes within the organization that do not result in better return on investment and higher stock prices quickly become low-priority initiatives. Just as the employees want their paycheck promptly on Friday and the customer expects the product or service delivered as scheduled, management's paycheck and the investors' return on their investment is driven by the quarterly results. As difficult as it would be for the employees and customers to be put off with a statement like: "You won't get it now; you have to look at the long-term results," it is equally as hard to put off executive management and the investor based on projected long-term results. As much as everyone agrees that organizations, as well as individuals, need to focus on long-term results, there is a basic need to satisfy today's needs as well by focusing on long-term changes to the culture of the organization. Keeping this in mind, the following are typical things that could serve as key improvement goals measurements for a PAM project:

- Customer satisfaction—customer-related measurement
- Value-added per employee—investor-related measurement
- Response time—customer-related measurement
- Return on investment from the improvement process—management-related measurement
- Morale index—employee-related measurement
- Dollars saved—management-related measurement
- Market share—investor-related measurement
- Error-free performance—management-related measurement
- Supplier index—supplier-related measurement

We suggest that the Steering Committee select four to seven key performance measurements around which the PAM process will be designed. The timing and selection of the transformation tools and when they are phased into the PAM five-year plan will be greatly impacted by how aggressive these accelerated performance goals are. We like to design the measurement system so that at least one of the key performance measurements is directly related to the employees, management, customers, investors, and suppliers.

> Goals. There's no telling what you can do when you get inspired by them. There's no telling what you can do when you believe in them. There's no telling what will happen when you act upon them.[1]
>
> **Jim Rohn**
> *Excerpts from The Treasury of Quotes*

Activity Five: Define Present Performance Levels of the Key Performance Measurements

In most cases, the Steering Committee will not have the information available at this point in the cycle on how all of the selected key performance measurements are operating at the present time. In this case, they will appoint a small committee to define the present performance level for each of these measurements and complete Activities Six and Seven.

Activity Six: Define Competitors' Present Performance Level for the Key Performance Measurements

In order for an organization to compete in today's international market, it is essential that it understands how its competitors are performing in these key performance measurement areas. For some of the measurements, it is better to use best practice data rather than limiting the organization to how their competitors are performing. One of the best ways to obtain this information is through conducting benchmark studies. A benchmark is defined as a "reference point for which other items can be compared." We are not recommending that the organization perform a benchmarking study at this time. Benchmarking is when an organization makes a study to determine which organizations are performing at the highest level related to the item being studied. It then contacts the best organizations to determine how they are functioning in order to get these results. The organization conducting the study then adopts or adapts these approaches to their process in order to bring about significant improvement. Benchmarking may be a tool that will be used later on in the PAM project, but at this time we only need a point of reference.

Activity Seven: Project Competitors' Performance Level for the Key Performance Measurements

Understanding the performance level of your competitors is only the starting point. If we designed our PAM process to be only as good as or a little better than our competitors, our organization would soon be out of business. Why? Because the competition is continuously improving and any organization that stops improving, no matter how good they are now, is not just standing still, it is losing ground because the competition is continuously advancing. As a result, once we determine the present performance levels of our competition or the best practice organizations, we need to project how they will be improving over the coming five years. One of the best ways to make these projections is to study their past performance and make the basic assumption that their future improvements will be equally as good or better than they were able to do in the past. These projected performance levels provide important input to setting the yearly goals for the PAM process. We accept the fact that it is impossible to make accurate five-year projections related to best practices and competitors' performance. The accuracy of any projection of this nature is dependent on the market development and it is only as good as the capabilities of the individuals making the projection. It is for this reason that it's important to select individuals to make these projections who are knowledgeable about the area being measured, thereby maximizing the confidence management can place on these projections.

Activity Eight: Finalize the Key Performance Goals for the Next Five Years

Prior to that point in time when the Steering Committee will be integrating the individual KPDs transformation plans into a Combined PAM Plan, the subcommittee conducting Activities Five, Six, and Seven should present the results of their activities to the Steering Committee. Based on the results of these activities, the business needs, and their own insight, the Steering Committee will finalize the key performance measurements and goals for the next five years. Even with all this effort, these goals are not cast in concrete. In today's fast-moving world, change occurs continuously and, therefore, there is a constant need for readjusting the organization's performance goals to meet these changing environments.

In order to prepare a final list of key performance measurements and goals, the Steering Committee will develop improvement in goals for each of the measurements. To accomplish this, they will need to define how the measurement will be made. For example, if the Steering Committee decides that customer satisfaction is one of the critical measurements, it could be measured by a survey administered to the external customers and using a 1 to 10 scale (1 being poor and 10 being outstanding). The Steering Committee could decide that they wanted a higher percentage of their customers to rate them between 8 and 10. The Steering Committee then needs to define, by year, a goal for each improvement measurement. Using the same customer satisfaction example, if the Steering Committee knew that 60% of their customers rated the organization between 8 and 10 today, their goal for customer satisfaction might be as seen in Table 4.1.

If the organization does not know the customer satisfaction level at the present time, they could set a target of reducing the number of customers that rate them below 8 (Table 4.2).

TABLE 4.1

Customer Rating Between 8 and 10

Year	Percentage Rating (%)
0	60
1	65
2	75
3	80
4	85
5	95

TABLE 4.2

Reduction in Ratings Below 8

Year	Percentage (%) Reduction in Ratings Below 8
1	10
2	Another 10
3	Another 20
4	Another 40
5	Another 40
Min. 5 year reduction	85

TABLE 4.3

Typical High-Level Set of Improvement Goals

	Year					
Measurement	**0**	**1**	**2**	**3**	**4**	**5**
1. Customer Satisfaction	60%	65%	75%	80%	85%	95%
2. Return on Investment from TIM		1:1	4:1	15:1	30:1	40:1
3. Average team return-on-investment		2:1		5:1		10:1
4. Defect rate improvement		2X		10X		100X
5. Value added/employee in $1000	45	50	65	70	75	80
6. New product cycle time in months	53	53		30		15

The same type of approach can be used for each of the other measurement goals. Table 4.3 is a set of typical improvement performance goals. It is easy to see that the PAM process must not focus only on meeting the vision statements, but also on meeting the yearly performance goals.

Table 4.3 is a set of goals we developed for one of our client's organization.

> Profits make it possible for us to grow. That expansion gives people the chance to develop their skills and advance professionally—in short, to stretch to their maximum potential. Growth enables us to take pride in belonging to a successful and well-respected company.
>
> **John Young**
> *CEO*
> *Hewlett-Packard*

SUMMARY

The first objective of the PAM approach is to define a clear picture (vision) of how the organization needs to change over the coming five years. In this chapter, the objective was to define how the transformation progress would be evaluated and how rapid these changes would be brought about. It is essential that the organization have a crystal clear view of where it's going and some way of measuring its progress to determine if they have their foot pressing on the accelerator at the right pressure.

Goals that are easy to reach provide little satisfaction. Goals that are impossible to reach result in mistrust and discouragement. But, stretch goals provide the challenge that makes us think in different ways, looking at things from different directions, and causing us to do the impossible.

H. James Harrington

REFERENCES

1. Rohn, J. 1994. *The treasury of quotes*. Lake Dallas, TX: Success Books.
2. Hackborn, D. 1990. *Business planning for competitive advantage: The ten step approach*. Palo Alto, CA: Hewlett-Packard Co.

5

Phase IV: Developing Individual KPD Transformation Plans

Define an improvement opportunity and then select the tools to take advantage of it. Too often organizations pick up a hammer and hunt for something to hit with it.

H. James Harrington

INTRODUCTION

During Phase IV, individual three-year transformation plans will be generated for some of the Key Performance Drivers (KPDs). We use the word *some* of the KPDs because often when education and training and measurements vision statements are used, they are included as part of the other individual three-year transformation plans. For example, training related to the job and measurement of the total process are key parts in this Business Process vision three-year transformation plan. These plans are designed to transform the organization from the AS IS state to the future state so that the future state is in line with the relevant KPD vision statements.

Note: You may question why we are talking about the three-year plans instead of just a five-year plan that is in line with the five-year vision statements. We have found that many of our clients, when they are just starting to prepare a long-range improvement plan, have a hard time defining the activities that they will be doing in year no. 4 and year no. 5. In these cases, we recommend that they start by only doing a three-year plan for the individual KPDs. When the individual KPD three-year transformation plans are combined together into a total plan, the additional workload imposed upon the organization will require that some of the individual transformation activities will need to be slipped to the fourth or fifth year. Many organizations have adopted this three-year planning approach to improvement in lieu of the piecemeal, "flavor-of-the-month" approaches to improvement that had been so unsuccessful for them in the past.

There are two different approaches that can be used to develop the individual three-year KPD transformation plans.

- One approach is to have the Steering Committee divide up into small subcommittees to develop draft copies of the plan. These subcommittees then present their draft of the plan to the Steering Committee. The Steering Committee reviews these plans and makes comments to the subcommittees, that then go back and reconfigure the plan based on the Steering Committee's inputs. The revised plan is then brought back to the Steering Committee for their approval. This is typically done at a two- or three-day offsite meeting.
- Another approach is to have the Steering Committee appoint subcommittees made up of subject matter experts and members of the Steering Committee to evaluate alternative plans and prepare draft plans over a one- to two-week period. A meeting of the Steering Committee is scheduled at which the subcommittees present their draft plans. Following the discussion with the Steering Committee, the subcommittees revise their plans. The revised plans are then presented to the Steering Committee for their approval.

After having used both of these two approaches, we find the second approach to be by far the most effective because it involves an expanded group of very knowledgeable subject matter experts who shared their knowledge and experience in selecting the methodologies and tools used in the individual KPD plans. This has the additional advantage of helping to get buy-in throughout the organization for the plans as they are rolled out.

During Phase IV, the Steering Committee will assign each of the KPD vision statements to a team that will be responsible for preparing a three-year transformation plan. (We will refer to these teams as subcommittees throughout this book.) This plan will define the tools/methodologies that will be used to bring about the required transformation as well as a projected timeline chart for the implementation of the tools/methodology. For this phase, the timeline chart should only state the number of months/weeks that is required for the various activities in implementing the tool/methodology. They also should suggest an individual or department to serve as the subcommittee leader and who will be responsible for coordinating the activities required to implement the tools/methodology. Phase IV consists of six activities:

- Activity One: Assign a planning team (subcommittee) to each KPD vision statement to develop an individual KPD transformation plan
- Activity Two: Define present-day problems
- Activity Three: Define roadblocks to evolving to the desired future-state vision
- Activity Four: Select tools/methodologies to address defined problems and roadblocks
- Activity Five: Develop an implementation timeline chart for each tool/methodology
- Activity Six: Obtain approval for the individual KPD transformation plans

Planning versus Problem Solving

Our experience indicates that North Americans and Europeans are good problem solvers, but they hate to do planning. Typically, there seems to be misunderstanding and confusion, but problem-solving activities are not part of the planning session. Planning sessions continuously flow over into problem-solving sessions. When planning sessions transform into problem-solving sessions (which the executive team loves to do), the meeting goes off in a separate direction, which slows things down to the point that the real planning activities never get completed. Planning is upper management's responsibility. Problem solving is the responsibility of middle managers, first-level managers, and the employees. Look at the difference between the two:

Planning	Problem Solving
Define direction	Define solutions
Identify change areas	Implement changes
Assign resources	Use resources
Identify needed action	Take action
Highlight symptoms	Find root causes
Look at big picture	Focus on single issues
Short-term cycle	Long-term cycle

Activity One: Assign a Planning Team (Subcommittee) to Each KPD Vision Statement to Develop an Individual Transformation Plan

There are many ways that this activity can be accomplished. Our preferred approach is to have the Steering Committee assign a team leader to pull together a subcommittee that includes subject matter experts and individuals who are knowledgeable about the problems related to the assigned vision statement. Often key individuals who will be responsible for implementing the plan also are included to get their early buy-in, knowledge, and support. For example, Table 5.1 shows typical subcommittee leader assignments by department for some of the individual KPD transformation plans.

The subcommittee leader assigned the responsibility to chair these subcommittees will be responsible for pulling together the appropriate subcommittee members to prepare each individual KPD transformation plan for the assigned vision statement. It is absolutely essential that at least one member of the subcommittee has a detailed knowledge of the tools/methodology required to address present-day problems related to the vision statement and the roadblocks that will need to be overcome to make the required transformation. Highly qualified, internal quality professionals, like Six Sigma Master Black Belts, will have the knowledge and experience to provide the subcommittee with an understanding of how the potential performance improvement tools and methodology operate and the type of problems they solve. If one is not available, it will be necessary to hire an experienced performance improvement consultant to provide this insight to the subcommittees. The subcommittees' activities should be boxed in so that all the subcommittees complete their assigned tasks within a maximum of 10 working days.

TABLE 5.1

Transportation Plan With Assigned Subcommittee Leader

KPD Transformation Plan	Typical Leader Assignment by Department
Management support/leadership	Personnel
Knowledge management	Technology
Customer interface	Marketing or sales
Supplier interface	Procurement
Training	Personnel
Process vision	Quality assurance or industrial engineering

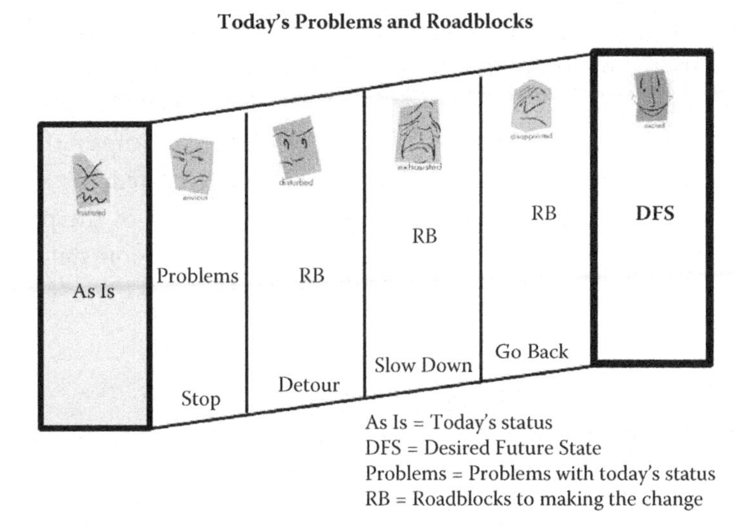

FIGURE 5.1
Present-day problems and roadblocks to making the required transformation.

One of the very first challenges that the subcommittees will face is to define the present-day problems related to the assigned vision statements and the roadblocks that will keep the personality/culture of the organization from making the required transformation. Figure 5.1 provides a visual presentation of the required transformation. The final individual KPD transformation plans that the subcommittee will be creating need to address not only today's problems, but the roadblocks that might stand in the way of a sustained major transformation within the organization.

Activity Two: Define Present-Day Problems

During the very first meeting of each subcommittee, a very detailed discussion related to understanding the vision statement needs to be held. It's very important that the subcommittee takes the time to thoroughly understand the intent of the vision statement and the transformation that is required to be in keeping with the statement. This will require that the subcommittee carefully dissect the assigned vision statement and discuss the way each word and phrase should impact how the organization functions and performs. The subcommittee also should study and understand the desired and undesired behavioral patterns and habits related to the assigned vision statement. By studying these desired and undesired behavioral patterns and habits, the subcommittee will gain additional understanding related to the transformation through which the organization is required to evolve.

Once there is a common understanding related to the impact the vision statement would have on the organization if implemented, the subcommittee is then in a position to discuss and make a list of the problems in the AS IS condition of the organization that the vision statement is designed to correct.

Activity Three: Define Roadblocks to Evolving to the Desired Future-State Vision

Definition

Roadblock: Something, such as a situation or condition, that prevents further progress toward an accomplishment.

It is relatively easy to define the problems related to the AS IS situation because the organization is presently living in this environment. It is far more difficult and requires far more insight to define the roadblocks that could occur that will detour and/or even prevent the transformation from happening. Much of this insight will come from analyzing the organization's past experiences in trying to modify its culture. Identifying these risks and applying risk management methodologies to mitigate these risks is absolutely essential for long-term success of the PAM process.

During this activity, each subcommittee will make a list of the risks/roadblocks related to making the required transformation from the AS IS state to the desired future state as defined in the assigned vision statement. Mitigation plans related to this list will be developed during Activities Four and Five.

Activity Four: Select Tools/Methodologies to Address Defined Problems and Roadblocks

Many subcommittee members, as they start this activity, feel that they are overwhelmed with the seemingly impossible task of making the desired transformation due to the major changes required in the organization's personality, the number of problems they need to solve related to the AS IS situation, and the major roadblocks that need to be overcome in order to make the required transformation. In addition, the subcommittee is often overwhelmed with the fact that there are literally hundreds of different performance improvement tools and methodologies available today and they are not sure how to choose the appropriate ones. (See Appendix B for a partial list of over 1,400 different performance improvement tools and methodologies.)

To add to all of this, many of the subcommittee members often realize how difficult it is to accelerate performance improvement in any organization. Even the individuals who were recognized as the gurus in the continuous improvement process, such as Philip B. Crosby, W. Edwards Deming, Armand V. Feigenbaum, Joseph Juran, and Kaoru Ishikawa (see Preface), could not agree on how an organization should implement the improvement process.

But the subcommittee members cannot let all of this overwhelm them. It only goes to reinforce the understanding that each organization has its own individual characteristics. It has its different culture, different management, different products, and a unique set of customers who each have their own unique set of customer requirements. Keeping this in mind, it is easy to understand why there is no one single approach that is right for all organizations. Every organization's improvement process has to be customized to its unique environment, culture, and performance needs.

Factors Impacting the KPD Transformation Plans

An organization must consider many factors before finalizing each of the KPD transformation plans. These considerations can be divided into two major categories: (1) impacting factors and (2) influencing data.

1. Impacting factors:
    a. Desired behavior and habit patterns
    b. Organization's mission
    c. Future business plans
    d. Organization's values
    e. Performance improvement goals
    f. Improvement/quality policies
2. Influencing data:
    a. Environment
    b. Opinion surveys
    c. Independent reviews

d. Improvement tools
e. Today's problems
f. Internal and external audits
g. Financial reports
h. Customer feedback data
i. Customer satisfaction reports
j. In-process performance data
k. Stock prices

Certainly the environmental influencing data can have a major impact on the final KPD transformation plans. Included in these considerations are items such as:

1. Technologies
2. Standards:
    a. State and federal government standards
    b. World body standards
    c. Self-imposed standards
    d. Government regulations
3. Desired pace of change:
    a. Changing customer expectations and needs
    b. New product releases to expand features and services
    c. Information technologies' impact on processes
4. Competitive environment

With so many methodologies, tools, and considerations, each organization has a wide selection from which to choose. It can be a blessing and a curse. It is a blessing because you can get just the right one for your situation. It is a curse because with so much to choose from, it is hard to make the best choice and you can actually choose to use too much. It is a lot like a boy going to a candy store where there is no one behind the counter. He just helps himself to all the candy he wants and goes out behind the barn to eat it. Soon he gets very sick and, based on this experience, he concludes that candy is not good for him (Figure 5.2).

We have seen presidents of organizations do the very same thing as the sick little boy who ate too much candy. They get disappointed in the improvement tools because they used too many of them at the same time. As a result, they believe that their chosen methodology is not good for their organization. Thus, the organization becomes ineffective because,

FIGURE 5.2
Boy in a candy store.

in addition to continuing its normal activities, it started too many new things at the same time without the resources to get them done, driving the organization into future shock, which caused the organization's performance to decrease rather than improve. This causes the organization to fail to realize the benefits that it could have received from the improvement activities.

> Improvement success is not measured by the number of improvements you undertake, but how well you implement the ones you commit to.
>
> **H. James Harrington**

Definition

Transition is defined as an orderly progress from one state, condition, or action to another.

An effective plan for transformation has the following characteristics:

- It accomplishes the desired results without rework.
- It is not abrupt.
- It does not create customer complaints.
- It does not create morale problems.

- It does not have scheduled slippages.
- It is not uncontrolled.
- It is not an unplanned occurrence.

By now the Steering Committee should realize that their assigned task is not an easy one. It's not as simple as copying something out of this book and putting their organization's name on it. It's a task that requires a great deal of technical excellence, understanding of the organization, and excellent analytic abilities. Figure 5.3 shows some of the considerations that influence and impact the individual KPD transformation plan design.

Typically, an organization would choose to develop vision statements for the following KPDs, so we will use the following to explain the process:

1. Manufacturing processes
2. Business processes
3. Customer/consumer partnerships
4. Management support/leadership
5. Supplier partnerships
6. Total quality management system
7. Measurement systems
8. Training

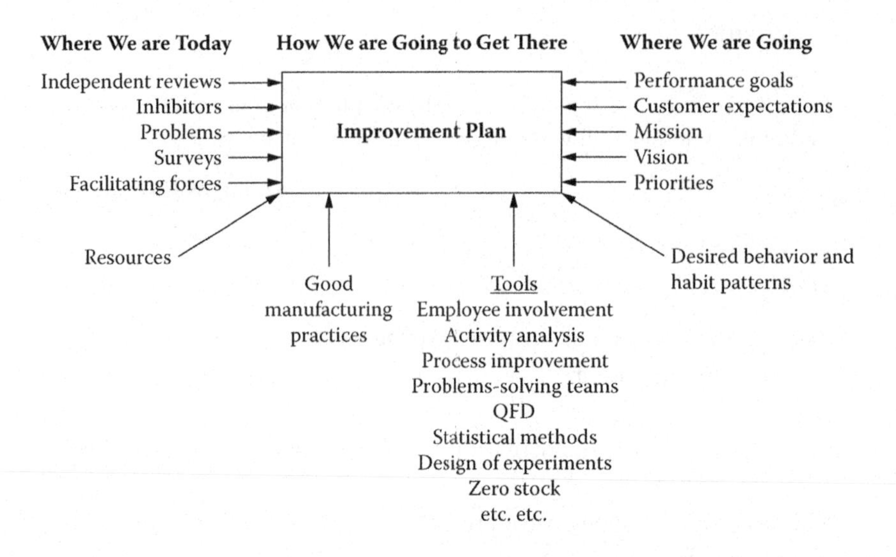

FIGURE 5.3
Considerations in defining an individual transformation plan for each KPD vision statement.

Note: We will assume that #7: Measurement Systems and #8: Training will be made an integral, individual KPD transformation plan.

As we have already pointed out, there are literally hundreds of different tools and methodologies that can be used to bring about performance improvement within an organization. The major task that a subcommittee now has is to sort through these numerous improvement tools and methodologies to select the best combination to attack the prepared list of problems and the identified roadblocks. The following are lists of the most common tools/techniques/methodologies that are used to bring about the transformation in some of the typical vision statements.

Frequently Used Tools for Manufacturing Process Vision Statement

- 5S
- Activity-based costing
- Concurrent engineering
- Design of experiments
- Design review
- Failure analysis
- ISO 14,000 or more stringent environmental requirements
- ISO 9000 or more stringent quality requirements
- Job training/certification
- Lean
- Management by walking around
- Manufacturability assessment
- MR P-2
- Organizational Change Management
- Problem-solving techniques, such as Plan-Do-Check-Act (PDCA)
- Process analysis and improvement
- Process qualification
- Process redesign
- Product cycle time controls
- Quality function deployment
- Reliability-centered maintenance
- Six Sigma
- Standardization
- Statistical process control
- Supply chain management

- Total Quality Management (TQM)
- Zero defects
- Zero stock: Just-in-Time (JIT)

Frequently Used Tools for Business Process Improvement Vision Statement

- Activity-based costing
- Area activity analysis
- Benchmarking
- Bureaucracy elimination
- Business process improvement
- Design for Six Sigma
- Fast Action Solution Teams (FAST)
- Organizational Change Management
- Process improvement teams
- Process redesign
- Process reengineering
- Simulation modeling
- Streamlined process improvement
- Task teams
- Value chain mapping

Frequently Used Tools for Customer/Consumer Partnership Vision Statement

- Area activity analysis
- Benchmarking
- Concurrent engineering
- CRM (customer relations management)
- Customer interface training
- Customer involvement in design reviews
- Customer phone calls (management and employees)
- Customer roundtables
- Customer surveys
- Customer visits
- Customer/relation measurements
- Design for Six Sigma
- Employee training

- Employee visits
- Empowerment of interface employees
- Improve call center reaction
- Innovation concepts
- Members of the design team
- Quality function deployment
- Simulation modeling
- Streamlined process improvement
- Supply chain management
- Total Quality Management (TQM)

Frequently Used Tools for Management Support/ Leadership Vision Statement

- Area activity analysis
- Behavioral modifications (executives, management, and team leaders)
- Building trust and understanding
- Competency evaluations
- Corporate governance
- Delegation/self-managed work teams
- Documenting and living up to the organization and personal values and principles
- Dual ladder in job descriptions
- Employee involvement
- Employee opinion surveys
- Four-way communications
- Innovation concepts
- Involving organized labor
- Job descriptions that define leadership responsibilities
- Leadership style:
    - Communications
    - Listening
    - Inspiring
    - Role models
    - Honesty
    - Leading not directing
    - Variables style interface based on employees' needs
    - Nonverbal communications

- Management by walking around
- Measuring executive error rate
- No layoff policies
- Organizational alignment
- Organizational Change Management
- Organizational master plan
- Performance planning and appraisal
- Performances toleration management
- Preparing visions and goals
- Rewards and recognition system
- Skip level interviews
- Suggestion systems
- Team building
- Time management
- Town hall meetings

Frequently Used Tools for Supplier Partnership Vision Statement

- Commodity teams
- Cost focus, not price
- Incentive plans
- ISO 14,000 or more stringent environmental requirements
- ISO 9000 or more stringent quality systems requirements
- Lot tracking during supplier production IT software
- Source inspection
- Statistical methods
- Supplier performance ratings
- Supplier qualification
- Supplier seminars
- Supplier surveys
- Supply chain management
- Supply management process (SMP)

Frequently Used Tools for Total Quality Management System Vision Statement

- Benchmarking
- Employee surveys
- Flattening the organization

- Four-way communications
- Internal audits
- ISO 14,000 or more stringent environmental requirements
- ISO 9000 or more stringent quality system requirements
- Lean Six Sigma
- Management audits
- Organizational Change Management
- Poor quality costs
- Quality assurance systems
- Quality planning
- Quality policy deployment
- Six Sigma
- Total Quality Management

There are seven basic principles that must be considered in developing any performance improvement plan. They include:

1. An organization must maintain long-term product/service superiority in its home market because this is the area that is most critical to profitability and growth. Organizations around the world have found out that if competition is possible, it's bound to occur. The only way to combat this competition is to anticipate it so you can stay ahead of it. It is difficult, if not futile, to play catch up.
2. Past experience has proved that foreign competition invades by establishing a small niche market, then expanding to capture bigger and bigger share of the total market.
3. Eliminating waste is more effective in increasing profits than increasing sales. In some cases, the reduction of waste has increased profits, more than doubling sales would have.
4. The soft side of customer relationships has proved to be more important than meeting specifications. Customer service must be surprisingly good to keep customers. Most companies can meet requirements. The truly great companies get ahead by meeting and exceeding customer expectations, then helping the customer set new, higher expectations for the total market.
5. Product leadership is not based on making products quicker and cheaper, but rather on making them better. A focus on "better" is the best way to produce less expensive products faster as well as sell them faster.

6. The best advertising a customer can have consists of groups of satisfied customers. A satisfied customer tells eight people, but a dissatisfied customer tells 22 people.
7. Technology can no longer be looked to as a wall that protects your customer base. Where product technology once provided a company with a competitive advantage, it now travels so fast from organization to organization and nation to nation that it has lost most of its advantage. The most important and powerful way to hold and increase your market share is through quality supremacy.

Assign a Knowledgeable Performance Improvement Specialist

Now is the point in time when the subcommittee needs to have someone who knows the different tools and methodologies that are designed to address the problems and roadblocks that the subcommittee has identified. Hopefully, there is some member of the subcommittee who can fulfill this role and provide this information to the other members of the subcommittee. If not, it will be necessary to go outside of the subcommittee and find someone who can fulfill this role. If the organization already has an internal Master Black Belt, as an employee, he or she should be capable of providing this type of assistance to the subcommittee. If the skills are not available within the organization, an outside experienced consultant will be required. This will be a valuable learning experience for the members of the subcommittee because they need to have a general understanding about a number of tools and methodologies and a good understanding of the tools they select to address the problems and roadblocks related to their vision statement. The task that the subcommittee must now address is to select the best ones from the potential tools/techniques/methodology that will be used to address the defined problems and roadblocks. In many cases, different tools can be effective at addressing the same problem. To make the selection even more complex, an individual tool, technique, methodology can address numerous problems and roadblocks. The subcommittee must study the individual tools, techniques, and methodologies to determine which of these tools provides the best combination of results in their particular environment. To accomplish this, a matrix (Table 5.2) provides a very effective approach.

Based on the example in Table 5.2, Tools 2, 3, and 4 would be the selected tools for this vision statement if there were no other considerations, but, of course, there are. We also have to consider the cost and risk related to

TABLE 5.2

Tools and Methodologies Versus Problems and Roadblocks

	1	2	3	4	5	6	7	Etc.
Problems								
1	×		×	×		×		
2		×		×	×			
3		×	×	×				
4			×		×			
5				×			×	
Etc.								
Roadblocks								
1	×			×				
2				×			×	
3		×				×		
Etc.								

implementing each of these tools. When these are taken into consideration, the subcommittee may find it necessary to look at more than these seven tools to find the best value solution for the total organization.

Another important consideration in selecting the tools that will be used to take advantage of the improvement opportunities is the fact that a single tool/methodology can have a positive impact upon other KPDs. Often an individual tool or methodology is selected and given priority based on its broad impact on the performance and cultural change across many KPDs. Appendix C provides a view of how some of the tools/methodologies that are used to take advantage of improvement opportunities in a single KPD can impact other KPDs.

Once the subcommittee has defined the tools/methodologies that will be used to support the transformation for the assigned vision statement, they should then classify each of the tools/methodologies into one of the following classifications:

- Should be implemented during the first year.
- Could be put off until the second year if there are not enough resources to do it during the first year.
- Could be put off to the third year if there are not enough resources to do it during the first or second year.

We realize that the subcommittee probably will want to have everything done during the first year, but we recommend that there should be at least one tool/methodology used in each of the three classifications. This provides for an ongoing emphasis toward reaching the vision statement. The subcommittee needs to understand that there will be a great demand for resources during the first year for each of the transformation plans for each of the vision statements and the organization must be careful that they don't start so many new activities that they will drive the total organization into "future shock." When organizations try to do too many changes at once, the organization will go into a state called *future shock*, which greatly decreases the probability of accomplishing the change activities and their normal activities.

Definition

Future shock is the point at which no more change can be accommodated without the display of dysfunctional behaviors within the organization affected. It is a term frequently used in Organizational Change Management Methodology.

Activity Five: Develop an Implementation Timeline Chart for Each Tool/Methodology

Once the subcommittee has selected the appropriate tools, a tactic-level implementation layout for each tool will be prepared and an individual will be assigned the responsibility for ensuring the plan is implemented. At this point in the planning process, the individual tool/methodology should be plotted on a Gantt chart based on the priorities defined in Activity Four. Figure 5.4 illustrates one page of a typical Gantt chart for an individual KPD transformation plan for a management leadership and support vision statement.

You will note that the timing is in months, such as 1, 2, 3, 4, 5, and so on, not in months of the year (January, February, March, etc.). This is done because the subcommittee is not responsible for scheduling the activity to be done in a specific month. The purpose of this chart is to show how long the project will take to complete and the interdependencies among activities. The following are the steps that a project will cycle through in order to implement a typical tool/methodology:

Key Change Focus Area <u>**Management Support/Leadership**</u>

Action = ▬ Ongoing Activity = //////////

Act. #	Activity	Month 1	2	3	4	5	6	7	8	9	10	11	12	Quarter 1	2	3	4	1	2	3	4	Person Responsible
1.0	Teams																					
1.1	Establish Task Team																					EIT/NCH
1.2	Establish Training Plan & Budget for ETT																					Task Team
1.3	Implement Training Plan																					Div. Pres.
2.0	Start Department Imp. Teams																					Dept. Mgr.
3.0	Develop Strategic Direction																					NCH
3.1	Communicate to Management																					
3.2	Communicate to Employees																					
4.0	Performance Plan & Appraisal																					
4.1	New Appraisal Process																					RJH
4.2	Communicate Plan to Management																					RJH
4.3	Communicate Plan to Employees																					Div. Pres.
4.4	Implement Plan																					Management
5.0	Measurement/Feedback																					
5.1	MBWA																					Div. Pres.
5.2	Employee Opinion Survey																					HJH
5.2.1	Feedback Results																					HJH
5.2.2	Re-survey																					HJH
6.0	Suggestion Systems																					
6.1	Establish Task Team																					NCH

X = Review R = Report P = Plan A = Approval
F = Form I = Implement T = Train

FIGURE. 5.4
Gantt chart for a KPD transformation plan for the management support/leadership vision statement.

1. Assign a project leader
2. Form and train the project team
3. Plan and prepare the materials for a pilot project
4. Conduct a project tollgate and obtain Steering Committee approval
5. Conduct the pilot project
6. Evaluate the pilot project results
7. Update the project material and plan for the follow-on to the pilot
8. Conduct a project tollgate and obtain Steering Committee approval
9. Conduct the follow up follow-on activities

Figure 5.4 is a computer-developed Gantt chart. In most cases, a software program like Microsoft Project® is used to document the KPD transformation plan. We used the computer-developed Gantt chart for the book because it was more legible when printed in book format.

The subcommittee should then review the performance improvement goals to identify which measurements are impacted by the specific KPD transformation plan item. The subcommittee should then evaluate how much improvement the specific change item will have related to the affected performance improvement goals. Although a number of the different tools and techniques can impact one measurement, the total sum of the impact does not necessarily add up to the total impact on the measurement because different tools may be claiming savings related to the same type of problem.

The subcommittee also should evaluate each of the specific transformation plans to ensure that they will be conducive to the desired behavior and habit patterns previously defined. If the individual transformation plan does not meet this test, it will need to be modified.

Activity Six: Obtain Approval of the Individual KPD Transformation Plan

During this activity, the subcommittee will present to the Steering Committee the individual KPD transformation plan for its assigned vision statement. This presentation should include the list of present-day problems and roadblocks to the transformation. It also should include a list of the tools that were considered and the reason why the subcommittee selected each of the individual tools. The subcommittee should present as well the timeline chart for each of the selected tools along with its assessment of the impact the tool will have on today's problems and future roadblocks.

They also will present an estimate of the resources required to implement the tools/methodology and the individual or department that should be responsible for coordinating the implementation of each tool/methodology.

Frequently, as a result of these meetings, a number of conditions surface that the subcommittee had not considered, requiring the subcommittee to call another meeting to further modify the recommendations. Often these Steering Committee recommendations are the result of the fact that it had the opportunity to review all of the individual KPD transformation plans and so it can see how they relate to each other. When this occurs, a second presentation is made by the subcommittee to the Steering Committee in order to obtain the Steering Committee's approval of the individual KPD transformation plan.

SUMMARY

During Phase IV, the Steering Committee assigned subcommittees to develop individual three-year transformation plans for each of the KPD vision statements. To accomplish this, the subcommittees had to first gain a thorough understanding of the meaning and intent of the assigned vision statement. The members then needed to conduct a gap analysis to understand the magnitude of the transformation that was required in order to be in compliance with the assigned vision statement. To develop the individual KPD transformation plans for each assigned vision statement, the subcommittees had to consider many factors. Some of them included:

- What is the desired behavior/habit patterns required to support the assigned vision statement?
- What are the present-day problems that have a negative impact upon performance related to the assigned vision statement?
- What roadblocks will the organization face in making the transformation?
- What kinds of resources are available to the subcommittees?
- Which combination of tools/methodologies would be the most cost-effective way to minimize the impact of today's problems and the future roadblocks for implementation of the KPD transformation plans?

- What impact would each of the KPD transformation plans have on the organization's performance?
- Will the proposed individual KPD transformation plan provide the organization with a competitive advantage?

It is very important that the subcommittee completes their assigned tasks in a very expeditious manner. During the PAM project, a lot of positive energy and momentum has penetrated the total organization as employees work with management to prepare a set of vision statements to define how the organization wanted to operate in the future. Typically, the employees are excited about what is going to happen in the near future. However, in too many organizations, the employees have been told by management that things are going to change and be better in the future only to be disappointed when management's promises fall short of meeting their expectations.

Employees listen to the tongue in their manager's mouth, but it is the tongue in their manager's shoes that really sets the culture in the company.

H. James Harrington

6

Phase V: Developing a Five-Year Combined PAM Plan

> Major plans must be based on and must consider all of the activities going on within the organization, not just what its objectives are.
>
> **H. James Harrington**

INTRODUCTION

During Phase IV, the subcommittees prepared individual transformation plans for the Key Performance Driver (KPD) vision statements. These plans were reviewed and approved by the Steering Committee. Now, reality has set in because there is a limited amount of resources that can be devoted to the Performance Acceleration Management (PAM) process. As much as the organization may want to improve and change, it still has to maintain its normal activities because today's processes are the ones that fund the desired transformations. It should be obvious to everyone by now that, although performance acceleration should free up resources, these resources will soon be absorbed based on the increased sales that should result from the PAM methodology. As a result, the organization will not be able to implement all of the positive changes that the subcommittee has identified during the first year. In truth, it could be very detrimental to the organization to overextend itself by taking on too many change projects at the same time. One of the biggest problems that we have faced in preparing PAM plans for our clients is that, by Phase V, the management team has become so committed to initiating these performance improvement approaches that they want to do them all at once, thereby overcommitting the organization. Taking on too many change activities at the same time is the biggest mistake an organization can make.

During Phase V, the Steering Committee, in conjunction with each KPD transformation plan subcommittee, will meet to establish a timeline chart that will combine the tools/methodologies required to bring about the accelerated performance improvement activities for all the vision statements. There are a number of important factors that need to be considered as the individual tools/methodologies are scheduled into the organization's activities. Some of these are typical considerations of the other projects that are going on within the organization, the timing related to the goals in the key performance indicators, the workload of the areas impacted by implementing the tools/methodology, fluctuations in workload at various times of the year, and skills available to implement the tools/methodologies. Phase V consists of five activities:

- Activity One: Define resource constraints
- Activity Two: Define interrelated tools/methodologies
- Activity Three: Prioritize individual tools/methodologies
- Activity Four: Prepare a Combined Work Breakdown Structure for the PAM Project
- Activity Five: Assign individuals or departments that will be responsible for the successful implementation of each of the tools/methodologies

Activity One: Define Resource Constraints

There are four types of activities competing for the resources within most organizations. They include:

1. The day-to-day activities that are defined by the department's mission statement
2. Personnel-related predictable resource constraints (absenteeism, holidays, vacations, training programs, etc.)
3. Periodic activities required to support the organization's operations, such as preparing budgets
4. Special projects

PAM projects fall into the fourth category: special projects. To effectively manage any organization, the first three activities must take priority over special projects. This presents a major problem that must be addressed when making any type of long-range plan. When we talk about resources related to scheduling the PAM process, we talk about it as it relates to people, machines, materials, and working capital. The rate that the PAM projects can be implemented is highly dependent upon the availability of these resources. In addition, implementing the PAM process has to compete with other special projects that are being implemented in the related/impacted parts of the organization. It is extremely important that you don't overcommit the available resources driving the organization into future shock where it could become nonfunctional. Fortunately, most organizations have some flexibility and options that they can use to help offset the additional workload that is imposed on the organization as a result of the special projects. Typical options that organizations use to provide additional temporary resources include:

- Outsourcing
- Loans
- Contracting for temporary help
- Consultants

Based on this knowledge, the first thing that needs to be understood before the timeline for the combined five-year PAM plan can be developed is to define how the organization has already committed to use its available resources, thereby defining the constraints related to implementing the combined PAM five-year plan.

There are many factors that can cause variation in the available resources. Some of them include:

- Vacations periods
- Holidays
- New product announcements
- Seasonal changes
- New equipment being installed
- Budget cycles
- New project proration and implementation
- Acquisitions
- Absenteeism
- Projected product demands

Understanding these variations in workload fluctuations in the individual functions and the organization as a whole is necessary to minimize the probability of sending individual areas into future shock. It is understood that any activity aimed at changing behavioral patterns and improving performance will increase the workload within the organization affected during the initial implementation. Once the behavioral patterns have changed, the efficiency of the individuals within the organization should increase significantly. The problem that management faces is how to handle the increased workload when a tool or new methodology is being developed and implemented within the organization. Too often in the past, management has ignored this problem and expected the individual areas to absorb the additional workload without increasing their resources. This is a good assumption if the individual areas were underutilizing their resources already, which is not the case in most organizations. When the organization makes the assumption that this additional workload can be absorbed with the present resources, it often means that the jobs the individuals (involved in the new projects) were doing will suffer. When employees reduce their attention and effort on their current assignments, this creates additional and new problems for the organization. One potential answer to the situation is to hire temporary help or to bring in consultants to aid in the implementation.

Activity Two: Define Interrelated Tools/Methodologies

Because the same individual tools/methodologies can be used to bring about improvement to a number of different KPDs, it is helpful to develop a matrix chart that shows what tools/methodologies are being used to bring about the transformation in all the KPDs' transformation plans (Table 6.1).

In Table 6.1, tool/methodology no. 2 is project management. You will note that project management is used in all eight KPDs' transformation plans. Tool/methodology no. 5 is Organizational Change Management and it is used in the KPD transformation plans for processes, customer, and management. This type of analysis helps the Steering Committee to understand which tools/methodologies will provide improvements across a number of KPDs. Often, this is a consideration that will allow the specific tool/methodology to be implemented earlier in the five-year Combined PAM Plan.

TABLE 6.1

Tools/Methodologies Used in the KPDs' Transformation Plans

Tools/Methodologies	1	2	3	4	5	6	7	8
1	×							
2	×	×	×	×	×	×	×	×
3	×		×					
4	×							
5		×	×			×		
6		×						
7		×						
8			×	×	×			
9			×					
10				×				
11					×			
12						×		
Etc.					×	×	×	×

Activity Three: Prioritize Individual Tools/Methodologies

The subcommittees have already prioritized their recommended tools/methodologies based on their importance to the individual KPD. When the Steering Committee evaluates all of the proposed tools/methodologies using the matrix developed in Activity Two of this phase, it is often necessary to shift some of the previously assigned priorities. In addition, the Steering Committee should look at the KPDs and determine the priority level of each driver.

Now, the most critical parts of the PAM process need to be completed. It is at this point in the process that the individual tools/methodologies are scheduled to be implemented within the organization. Many factors need to be considered in making these decisions. Some of the more obvious ones include:

- Availability of resources
- Performance objectives
- Competitive pressures
- New product/services development
- Organization's culture
- Performance level of the individual functions
- Level of management support
- Organization's level of support to change

In some organizations, the Steering Committee has decided that they would have a specific emphasis for each of the five years. For example:

- Year One: The focus could be on Management Support and Leadership.
- Year Two: The focus could be on Business Processes.
- Year Three: The focus could be on Customer Partnerships.
- Year Four: The focus could be on Supplier Improvements.
- Year Five: The focus could be on Innovation.

Having a theme for each year does not mean that there isn't work done on all of the KPDs. It only means that a major focus will be put on tools and methodologies that have the biggest impact on the theme for that year.

Activity Four: Combine the Individual KPD Transformation Plans into the PAM WBS

During Activity Four, the individual KPD transformation plans are combined together into a total project Work Breakdown Structure (WBS) creating a basic document for the Combined PAM Plan. To accomplish this, many factors need to be considered, including:

- Availability of resources
- The need to be profitable early in the improvement cycle
- The change history of the organization
- Performance improvement goals
- Priorities established by the Steering Committee
- New products being released
- Availability of equipment
- Money availability
- Other projects being implemented
- Impact of acquisitions
- Organization's products strategic plan

There are two different approaches that we have found effective in accomplishing this activity:

1. One approach is to have a committee set up to develop a straw man's version of the Combined PAM Plan that would be presented to the Steering Committee for approval. This committee would be made up of the leader of each of the subcommittees that developed the individual KPD transformation plans, plus a facilitator who would lead the meeting.
2. Another approach is to have the Steering Committee invite the leaders of each of the subcommittees to attend a meeting where the Combined PAM Plan would be prepared.

In the first approach, the Steering Committee combines together all of the tools and methodologies in the individual KPD transformation plans. They then define the top priority KPDs and schedule all of the activities that were classified as *should be implemented in the first year* by that subcommittee. If that included the top priority tool/technique from each of the subcommittees, the Steering Committee then sets a priority to the tools/techniques that will have the maximum benefits in multi-KPDs. Based on this, the workload in the first year would be analyzed to determine how it should be distributed throughout the year. If there are still resources available during that first year, other recommendations for implementation during the first year, which were not already scheduled, would be evaluated to define the best fit based on the skill resources that are available.

There are four activities that should be included in the first year's plan:

- Regularly scheduled program review meetings of the Steering Committee
- Communication plan related to the PAM methodology
- Plan on how excess resources will be deployed other than having layoffs
- Organizational Change Management activities

With this approach, the Steering Committee then focuses on the projects that would be conducted during the second year. In some instances, many of these projects that are started during year 1 could continue to absorb resources during the second year. On the other hand, some of the projects started in the first year often could result in making additional resources available during the second year because they eliminated no-value-added activities and increased productivity. (This use of resources must be first considered before new projects are added.) First priority during the second year should be given to all of the tools/methodologies that were rated as *should be implemented during the first year.* If there are still resources available after these projects have been added to the workload, then the tools/methodologies, which the subcommittee agreed could be put off to year 2, can be added to the WBS based on the priority that each of the KPDs were given by the Steering Committee in the previous activity. This approach is repeated for years 3, 4, and 5.

Next, the Steering Committee would go back and review the activities as scheduled on the WBS to determine if the performance improvement targets could be met. This often leads to some slight shifting of projects, moving some ahead and some back in the WBS.

Another approach that we have found to be very effective is to have the Steering Committee hold a meeting with the leaders of each subcommittee. At this meeting, the WBS is created by making a matrix of seven columns. In the left-hand column, you would list all of the tools/methodologies defined by the subcommittees (Table 6.2). The headings of the other columns are

- Column 2: Employee Impact
- Column 3: Dollars Saved
- Column 4: Business Impact
- Column 5: Customer Impacts
- Column 6: Total
- Column 7: Suggested Year to Implement

TABLE 6.2

Approach to Prioritizing Strategic Plans

Strategies, Tools and Methodologies	Employee Impact	Dollars Saved	Business Impact	Customer Impact	Total	Improvement Year
Lean	3	2	3	1	9	1
Knowledge Management System	4	2	4	2	12	3
Operating Manual	3	2	2	1	8	2
Team Training	4	2	3	1	10	1
Self-Management Teams	5	2	1	0	8	4
Organizational Alignment	4	3	5	2	14	1
Supply Chain Management	1	3	4	2	10	2

Then, the group would rate every block in the chart from 1 to 5 as follows:

- 1 = Very low impact
- 2 = Low impact
- 3 = Average impact
- 4 = High impact
- 5 = Very high impact

In this case, the higher the total, the bigger the impact on the organization. In most cases, the high-impact item was scheduled early in the cycle, but not in all cases. In some cases, an item cannot be implemented until other conditions are completed first. In Table 6.2, only seven tools/methodologies are listed to present a simple example. Typically the list would be 25 or more strategies, tools, and methodologies.

Preparing a Five-Year Work Breakdown Structure

The Steering Committee will now prepare a five-year WBS for the PAM process. This WBS will be used to show the activities related to the tools that are scheduled for implementation during the first five years. The WBS will be defined by months for the first year and by quarters for the next two years. The tools and methodologies plan for the fourth and fifth year usually will be listed just in the project plan, but not included in the WBS. The WBS should be updated every six months dropping off the previous six months' activities and adding the next six months to it. For example, at the first update, the six months of the first year will be dropped off and six months of the fourth year will be added. This approach is used because it is very hard to predict the many changes that can occur in an organization's environment over the next three years or how effectively the PAM process will be implemented over the same time period. Figure 6.1 is a typical example of one page of a WBS at the start of a PAM process.

You can see at this stage that we have taken into consideration things such as summer vacation, holidays, strategic planning cycles, and so on. A major portion of this plan was the supporting Organizational Change Management (OCM) plan that helped ensure the smooth implementation of the individual process's changes.

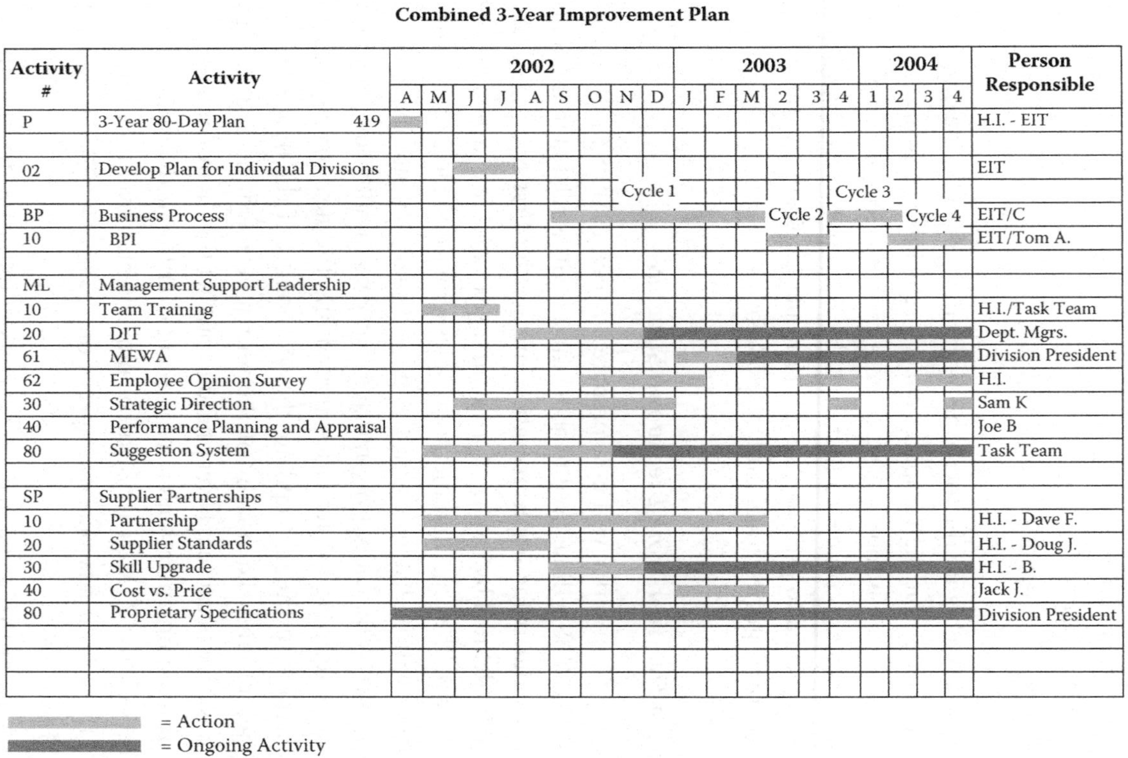

Combined 3-Year Improvement Plan

Activity #	Activity	2002									2003						2004				Person Responsible
		A	M	J	J	A	S	O	N	D	J	F	M	2	3	4	1	2	3	4	
P	3-Year 80-Day Plan 419																				H.I. - EIT
02	Develop Plan for Individual Divisions																				EIT
										Cycle 1							Cycle 3				
BP	Business Process													Cycle 2					Cycle 4		EIT/C
10	BPI																				EIT/Tom A.
ML	Management Support Leadership																				
10	Team Training																				H.I./Task Team
20	DIT																				Dept. Mgrs.
61	MEWA																				Division President
62	Employee Opinion Survey																				H.I.
30	Strategic Direction																				Sam K
40	Performance Planning and Appraisal																				Joe B
80	Suggestion System																				Task Team
SP	Supplier Partnerships																				
10	Partnership																				H.I. - Dave F.
20	Supplier Standards																				H.I. - Doug J.
30	Skill Upgrade																				H.I. - B.
40	Cost vs. Price																				Jack J.
80	Proprietary Specifications																				Division President

= Action
= Ongoing Activity

FIGURE 6.1
First three-year combined PAM WBS.

Note: Remember, the entire concept of change management needs to be well understood by the Steering Committee and the people who implement the plan because it plays an integral part in the behavior/habit transformation activities that are required to support and maintain the gains that the PAM methodology brings to the organization.

Figure 6.1 is a graphic display of the WBS. In most cases, a computer-generated WBS using a program like Microsoft Projects® is used to manage the projects. We use a graph in Figure 6.1 instead of a computer-generated version because it is easier to visualize in book format than the computer-generated WBS.

> Plans are only good intentions unless they immediately degenerate into hard work.
>
> **Peter F. Drucker**
> *American author*

SUMMARY

During Phase V, the individual KPD transformation plans were combined into one consolidated five-year plan designed to accomplish the desired transformation. To accomplish this, many competing activities and variations in availability of resources had to be considered. It is particularly important to curb the Steering Committee's enthusiasm related to the PAM methodology so that they do not overcommit the organization to implementing too many changes at one time. Being overly aggressive in this is a sure formula for failure. Implementing a change within an organization imposes a level of emotional stress on its employees. When the stress of individual changes going on in an area accumulates, the total organization can go into future shock. When future shock occurs, the organization becomes dysfunctional. It is for this very reason that an effective OCM approach needs to accompany the PAM methodology implementation.

> Managers who are overly enthusiastic about change are a bigger problem than managers who resist change.
>
> **H. James Harrington**

7

Phase VI: Implementing the Combined PAM Plan

The best plan poorly implemented is just an exercise in futility. It sets expectations and results in disappointment and lack of trust in management.

H. James Harrington

INTRODUCTION

During Phase VI, the Combined PAM (Performance Acceleration Management) Plan will be implemented. Based on the schedule that was defined in the Work Breakdown Structure (WBS), which was developed in Phase V, the individual or department who was assigned the responsibility for implementing a tool/methodology during the first 90 days of the project will prepare a detailed implementation plan. These detailed, individual Key Performance Driver (KPD) transformation plans will be integrated together into a Rolling 90-day WBS. Every 30 days an additional 30 days will be added to the Rolling 90-day WBS. These additions will reflect the progress made (or lack of progress) in the previous 30 days, plus the additional tools/methodologies that were scheduled in the five-year plan to start during the next 30 days. This plan will include critical tollgates that will ensure that the project is on schedule and that the expected results and goals will be met. Phase VI consists of six activities:

- Activity One: Develop individual detailed implementation plans for each tool/methodology.

- Activity Two: Combine the individual detailed implementation plans into a Rolling 90-day WBS.
- Activity Three: Prepare a three-year financial plan to fund the PAM project.
- Activity Four: Establish the tracking system to ensure the project is on schedule, within costs, and will produce the desired results.
- Activity Five: Establish a measurement system that would measure the impact the project is having on the organization's performance.
- Activity Six: Evaluate contributions made by individuals, groups, and teams, and recognize outstanding performance.

Activity One: Develop Individual Detailed Implementation Plans for Each Tool/Methodology

Now each of the subcommittees has a responsibility to generate a detailed implementation plan for all of the tools/methodologies that were approved for implementation during the first year of the project. Each line item on the detailed implementation plans should have a start and end date, plus the name of the individual who or department that is responsible for completing the activity. Close attention should be paid to ensure that activities requiring input from another activity are not scheduled until the interrelated input is available. Careful analysis should be made of each activity to ensure that the activity is assigned to an individual. Often an individual assigned to complete an activity will require training related to the activity. In these cases, part of the detailed implementation plan should include time for training the individual so that the activity can be performed in a more satisfactory way.

The layout for these detailed implementation plans will be very similar to the one that was originally prepared by the subcommittee for the individual KPD transformation plans, with the exception that specific dates defined in the Combined PAM Plan will be used only in scheduling the tools/methodology that were approved for use during the first year of the project. In addition, the activities required to implement the tools/methodology will be defined in detail along with the start and completion date for each activity and the individual or department that is responsible for the activity.

Table 7.1 is a typical example of how a detailed implementation plan might look for installing a suggestion system.

TABLE 7.1

Typical Detailed Implementation Plan for Installing a Suggestion System

1.0	Establishes the suggestion process development team: May 7–May 11: Tom Jenkins
2.0	Benchmark suggestion systems
2.1	Collect public domain information: May 11–May 25: Mary Wilson
2.2	Defined organizations to benchmark: May 25–June 11: Suggestion development team
2.3	Conduct discussions with target organizations: June 1-June 15: Suggestion development team
3.0	Prepare operating procedures: June 1–June 25: Thomas Lee
4.0	Obtain management approval: June 20–June 27: Tom Jenkins
5.0	Install suggestion system: June 25–July 7: Thomas Lee
6.0	Communicate suggestion system to total organization: July 10–July 15: Mary Wilson
7.0	Measure results: July 15–October 1: Thomas Lee
8.0	Upgrade suggestion system based on experience: July 25–October 15: Suggestion development team
9.0	Maintain suggestion system: Ongoing: Personnel department

The subcommittee also will prepare estimates of the resources required to support the development and implementation of each of the tools/methodologies that was approved for the first year of the project.

Activity Two: Combine the Individual Detailed Implementation Plans into a Rolling 90-Day WBS

Now the individual detailed implementation plans will be combined into a total WBS for the first year of the project. Figure 7.1 is a typical page and graphical display from a Rolling 90-day Combined WBS.

In most cases, a computer-generated WBS using a program like Microsoft Project® is used to manage the projects. We use the graph in Figure 7.1 instead of a computer-generated version because it is easier to visualize in book format than the computer-generated WBS.

At four-week intervals, the previous four weeks will be removed from the Rolling 90-day Combined WBS and the next four weeks will be added. Adjustments will be made to the WBS that reflects any changes or slips that occurred over the past four weeks. Any slips in commitments should be reviewed and approved by the Steering Committee. Using this approach,

Action = ▬
Ongoing Activity = //////////

Act. #	Activity/Person(s) Responsible	Week Ending Dates													Person(s) Responsible
P	Planning	4/5	4/12	4/19	4/20	5/3	5/10	5/17	5/24	5/31	6/7	6/14	6/21	6/28	
1.0	3-Year/90-Day Plan														
1.1	Transcribe EIT Written Plan														E&Y/KL
1.2	Draft 90-Day Plan														E&Y/KL
1.3	Mail Draft to EIT and Division Czars														E&Y/KL
1.4	EIT Modifies/Approves Plan														EIT/NCH
1.5	Return Revised Plan to E&Y														EIT/NCH
1.6	Revise Plan as Needed														E&Y/KL
1.7	Present Final Plan to EIT			X		A	I								E&Y/KL
2.0	Develop Industrial 3-Year Plans														E&Y/DIV.CZAR
ML	Management Support/Leadership														
1.0	Teams														
1.1	Establish Task Team														EIT/NCH
1.1.1	Identify Needed Training														Task Team
1.1.2	Develop Training Plan					A									Task Team
1.1.3	Establish Training Budget						R								Task Team
1.1.4	Present Budget and Plan to EIT							A							Task Team
1.3	Implement Training Plan								I						DIV. Pres.
3.0	Strategic Direction														NCH
4.0	Performance Plan and Appraisal														
4.1	New Appraisal Process								R						RAH
6.0	Suggestion System														
6.1	Establish Task Team														EIT/NCH
6.2	Determine Type of Suggestion System														Task Team

X = Review R = Report P = Plan A = Approval
F = Form I = Implement T = Train D = Done

FIGURE 7.1
Typical Rolling 90-day Combined WBS.

the organization will always have an up-to-date, accurate plan that defines the activities that will take place in the PAM implementation plan.

Activity Three: Prepare a Three-Year Financial Plan to Fund the PAM Project

Every organization is concerned about how it spends its limited resources. It is obvious that the PAM project will require the involvement and time of many of the organization's employees. In addition, it may be necessary to acquire outside help to handle the additional resources and to provide knowledge and experience related to the use and implementation of specific tools/methodologies. The organization is particularly vulnerable to requiring additional resources during the first year of the project. The performance improvement gains during the first year should more than make up for the additional cost related to implementing the project during years 2, 3, 4, and 5.

As a result, the major focus related to the use of resources is directed at year 1 of the project. During Activity Two of Phase VI, the subcommittees prepare estimates of the amount of resources and the type of resources that would be required to implement each of the tools/methodology that were approved for implementation during the first year. In most organizations, this is a combination of software, hardware, facilities, staff, and consulting services. Frequently, the individual coordinating a subcommittee requires resources over which he/she does not have direct control. As a result, we recommend that someone from Finance be assigned to pull together all of these resource requirements into a single package broken out by the functional managers that will need to provide the resources. After the individual functional managers review these inputs, the affected functional managers will submit a request to have their budget and headcount requirements modified. Any request for increased budgets or headcount should be closely reviewed by the appropriate members of the executive team. Request for increases in budget and headcount during the first year also should be accompanied by decreasing budget and headcount in the next two years that will more than compensate for the increases they require during the first year.

Activity Four: Establish the Tracking System to Ensure the Project Is on Schedule, within Costs, and Will Produce the Desired Results

Implementing the PAM methodology represents a major undertaking for the organization. PAM projects can be important and provide bigger returns over a long period of time, often resulting in a new, successful product. It is a major part of an organization's Strategic Plan and its activities should be handled in a very professional manner. If the organization has a Project Office, this is the department that should be assigned to monitor the implementation of the methodology. Each of the major tools/methodologies should be treated as a project unto itself with its own project charter and project accountability. If the organization does not have a Project Office, the Steering Committee should assign an individual the responsibility of coordinating, measuring progress, holding phase reviews/tollgates, and immediately informing the appropriate members of the Steering Committee when exposures to project slippage or failures are identified. During the first year of the project, the Steering Committee should meet a minimum of once a month to conduct a complete review of the portfolio of projects that make up their PAM initiative.

In addition, the individual executive whose employee has been assigned to head up the implementation of a tool/methodology also is held accountable for the successful implementation of that tool/methodology. As a result, the executive also will serve as a mentor and an advisor to the individual in charge of implementing the tool/methodology.

Activity Five: Establish a Measurement System That Will Measure the Impact the Project Is Having on the Organization's Performance

> In most organizations, a measurement system that is related to its KPDs needs to be developed.
>
> **H. James Harrington**

A good measurement system is an absolute essential ingredient for any improvement process. It needs to be established early in the cycle so that the starting point of the improvement effort is effectively measured and documented. During Phase IV, the subcommittees selected tools/methodologies based on the problems the organization was having and identified the roadblocks to making the required behavior/cultural changes. The tools/methodologies selected were designed to bring about positive shifts in the way the organization operates. They may be designed to reduce costs, improve quality, improve employee morale, reduce cycle time, reduce inventory, increase turnover rate of inventory, increase customer satisfaction, etc. In each of these cases, they are designed to make a tangible improvement in the way the organization is performing. The individual or team responsible for implementing the tool/methodology in this project has a responsibility for defining how the impact of this tool/methodology will be measured. These individuals, in relation to the measurement system of the tools/methodology they are implementing, will be responsible for:

- Defining what will be measured
- Defining how it will be measured
- Defining when it will be measured
- Defining who will do the measurements
- Defining how the data will be analyzed
- Defining the target improvement level
- Getting the measurement system implemented
- Ensuring the data collected are meaningful
- Analyzing the data
- Reporting the results

As you can see, establishing an effective measurement system is a major part of the activities performed by the team that is responsible for implementing the tools/methodologies.

Activity Six: Evaluate Contributions Made by Individuals, Groups, and Teams, and Recognize Outstanding Performance

> No bad deed goes unrecognized, but too many outstanding accomplishments are never recognized.
>
> **H. James Harrington**

As the individual projects near completion, it is time to consider the contributions made by the individuals involved in the project, and the team or groups that worked together to design and implement the project. In some organizations, just to complete a project is a major accomplishment. In other organizations, the success of a project is based on the amount of money that was saved. Another organization may measure success based on the project meeting or exceeding the performance improvement goals that were agreed to at the beginning of the project. Still another organization will recognize individuals based on the amount of effort and overtime they dedicated to completing a project. Many organizations will recognize individuals and teams based on how creative they were in solving the problem. The truth of the matter is, we recognize individuals as doing a satisfactory job when they receive their regular paycheck. However, that doesn't mean that management should not do something special to recognize the team when a project is completed. The type and amount of the recognition should vary based on a number of considerations. Some points that should be considered include:

- The difficulties faced to complete the project.
- The creative way the project was handled.
- The creative solution that was developed.
- The effort that was needed to complete the project.
- Conformance to budget and schedule.
- The impact the project will have on the organization's performance.

We do not recommend that a high emphasis be placed on the amount of money saved as a result of implementing the tool/methodology. Many projects by their nature can provide large return on investment without requiring a great deal of effort and creativity, while other projects may be very difficult to complete and their outcome is measured in less tangible things like improved employee morale, increased customer satisfaction, reduced cycle time, etc. Typical team recognition/award activities include:

- A simple "thank you" for a job well done given at a Steering Committee meeting.
- Pizza provided by the company for everyone at the last meeting of the subcommittee.
- An article about the subcommittees' activities in the organization's newsletter, accompanied by photographs of the subcommittees.
- A picnic for the members of the subcommittee and their families.
- Luncheons with upper management.
- Subcommittee attendance at a technical conference.

- Group mementos (pen sets, calculators, merchandise, T-shirts, etc.).
- Time off from work.

Many organizations today only focus on team recognition. We believe that it is important to recognize an individual's contribution to developing creative solutions, managing the project, individual effort, and commitment to the successful completion of the project. We personally believe there are significant advantages to recognizing individual contributions as well as team contributions. On most teams, there are individuals who do most of the work and contribute most of the ideas while other individuals just float along doing minimum work. In some cases, certain team members may be required to contribute to just a very small part of the project's activities due to his/her limited personal skill set. Based on these and other factors, individuals who are performing in an outstanding manner need to be recognized and rewarded. However, recognizing individuals can often create problems and jealousies between members. For example, if one member gets special recognition, it might make another member unhappy because, from the latter's standpoint, he or she contributed as much to the success of the project as the individual team member who got recognized. One way to counteract this problem is to ask the team to identify individuals who did an outstanding job in making the project successful. One of the most significant rewards for any employee is to receive the sincere appreciation of his or her peers for a job well done. Frequently, these types of awards are associated with professional societies. Today, the same concept has become a very positive motivating force within many organizations, where the award recipients are selected by the employees, not by management.

Often, management establishes the basic ground rules and financial constraints related to peer recognition awards. The employees should then define what type of behavior should be recognized and establish how these behaviors will be rewarded. Empowering the employees to select the reward frequently provides management with very pleasant surprises related to the creative way employees apply the limited reward budget. Our experience indicates that employees usually plan something that is fun and the winners are usually very touched. Typically when management prepares an award ceremony, it turns out to be a ceremony. When employees plan the same event, it turns out to be a party, and the awards that are presented turn out to be from the heart rather than from the pocketbook.

We like to provide the team with some guidelines related to the percentage of the team members that might be eligible for special recognition.

Typically no more than 25% of the team members would be selected for special recognition. If the results of the team's activity are outstanding, the total team should be recognized in some manner as well as selecting individuals for special recognition. Often there is no one or two individuals on a team who have had significant impact on the outcome of the team's activities, enough to be selected to receive special recognition. When this occurs, it's a good indication that the team was well motivated and managed.

Some typical ways to reward and recognize individual team members include:

- Promotions
- Trip to customer locations
- Attending the organization's recognition meetings
- Gifts of jewelry
- Special parking space
- Articles in the newsletter
- Employee's picture on a poster
- Employee of the month
- Public notice posted on the bulletin boards. (One plant in New York City has a huge billboard on the top of the plant that flashes the names and accomplishments of special employees.)
- Trophies
- Award pins
- Tickets to events
- Dinner or night out on the town for two
- Monetary awards

Everybody hears "thank you" in a different way. The management team has to be very creative in defining what type of recognition will reinforce the desired behaviors without creating jealousy throughout the organization. To stimulate your thinking, we suggest that you read *1001 Ways to Reward Employees*[1] by Bob Nelson (Workman Publishing Company, Inc., 2005).

> Compensation is what you give people for doing the job they are hired to do. Recognition, on the other hand, celebrates an effort beyond the call of duty.[2]
>
> *Incentive Magazine*
> *How to Profit from Merchandise Incentives*

SUMMARY

Implementation is the pot of gold at the end of the rainbow.

H. James Harrington

In too many organizations the project loses luster when concepts are transformed into realities. In the initial stages of the accelerated improvement initiative there is a high degree of excitement, creativity, and challenges related to defining problems, identifying root causes, and developing creative solutions to difficult problems and improvement opportunities. There is even a degree of challenge and enjoyment in defining how the problem solution will be rolled out throughout the organization and estimating the degree of improvement and savings that will result from the project. However, now the real work begins as the organization transfers concepts into realities. People bring up things that no one had considered. It seems like everybody is trying to find what is wrong with the creative solution that the subcommittee developed instead of going about positively implementing the concepts. Up until this point, it was the members of the subcommittee working together in the office area or conference room with a common goal. Now, all of a sudden, the people that you are trying to help are fighting back and resisting the concepts and approaches that the subcommittee had worked so hard to create. It's often hard for the subcommittee members to believe that everyone wouldn't be as excited as they are about implementing the concepts. The subcommittee members think: "How could anyone question not implementing this brilliant solution we have created to the problems they are facing?" But remember, often the new processes developed by the subcommittee result in increased productivity which, from the standpoint of the people affected, could mean loss of their jobs. It is for this very reason that we recommend that all organizations establish a no layoff policy that occurs as a result of performance improvement initiatives. This does not mean that the organization should or will not have layoffs as a result of decreases in demand for the organization's output.

Yes, as the old saying goes, implementing the performance improvement solutions "is where the rubber meets the road." Resistance to change is just human nature. In fact the implementation team should be concerned if there isn't some resistance to implementing the solutions. The key to a successful implementation is to use an effective Organizational Change

Management methodology to minimize the magnitude and duration of the resistance to change. Careful consideration must be given to when, where, and how the changes are introduced into the organization.

During Phase VI, a great deal of emphasis is placed on establishing effective measurement systems, tracking the portfolio of projects that are underway within the organization, measuring the impact of the change on the organization, and, finally, recognizing individuals and teams whose performance is outstanding.

> The best idea is only as good as the way it is implemented.
>
> **H. James Harrington**

REFERENCES

1. Nelson, B. 2005. *1001 ways to reward employees.* New York: Workman Publishing Co.
2. Anonymous. 1991. How to profit from merchandise incentives. *Incentive Magazine* special supplement 165(9): 4–34. September.

8

Phase VII: Continuously Improving

When an organization stops improving, it starts slipping backwards.

H. James Harrington

INTRODUCTION

All too often an organization will implement initiatives that bring about very positive changes in its operations and output performance and all of a sudden take a deep breath and relax. No, that's wrong; they take a deep breath and collapse. They feel that they have succeeded because they reached their goals, wherein in truth, they have just reached one milestone in the evolution of their organization. Yes, it's the right thing to do to take time to celebrate the accomplishment and to recognize those who have contributed greatly to meeting these goals. But, as soon as goals have been reached, it is time to move the bar up and renew your effort to be even better. You have to feel good about being better today than you were yesterday, but you need to be better tomorrow than you are today. It's a lot like competition in sports. As soon as a record is set, you need to strive to beat it and be better. Every organization needs to strive to break its own records as well as the records set by its competition (Figure 8.1).

THE NEXT CYCLE OF PAM

Early in the fifth year of the Performance Acceleration Management (PAM) five-year cycle an organization should determine if its ongoing

FIGURE 8.1
Now the bar moves up.

improvement efforts should be based on recycling the PAM process (starting with Phase I) or if it should focus on a more standard approach to ongoing continuous improvement allowing the organization's culture to remain in a more stable state. Many organizations decide that they are ready to recycle the PAM process as they want to bring about additional changes to the KPDs. In these cases, it is important to start Phase 1 of the PAM process early enough in the fifth year so that the five-year Combined PAM Plan is completed prior to starting the Annual Operating Plan development cycle. This allows the new five-year Combined PAM Plan to become part of the budget for the total organization. The more advanced organizations also are now ready to develop an Organizational Master Plan.

Organizational Master Plan

Definitions

Organizational Master Plan: The combination and alignment of an organization's Business Plan, Strategic Business Plan, Combined PAM Plan, and Annual Operating Plan.

Business Plan: A formal statement of a set of business goals, the reason they are believed to be obtainable, and the plan for reaching these goals. It also contains background information about the organization.

Strategic Business Plan: This plan focuses on what the organization is going to do to grow its market share. It is designed to answer the questions: What do we do? How can we beat the competition? It is directed at the product and/or services that the organization provides as viewed by the outside world.

Combined PAM Plan: This plan focuses on how to change the culture of the organization. It is designed to answer the questions: How do we excel? How can we increase value to all stakeholders? It addresses how the organization's controllable factors (Key Performance Drivers (KPDs)) can be changed to improve the organization's reputation and performance. Some organizations call this their Strategic Improvement Plan.

Annual Operating Plan: A formal statement of the organization's short-range goals, the reason they are believed to be attainable, the plans for reaching these goals, and the funds approved for each part of the organization (budget). The Annual Operating Plan is often just referred to as the Operating Plan.

Many of the advanced organizations around the world are operating using an Organizational Master Plan that harmonizes together four different and unique plans (Figure 8.2).

The organization's Business Plan is a basic requirement for all newly formed companies. It contains the organization's mission, objectives, and value statements. It is primarily prepared to crisply define the organization's intent, the markets it wants to service, things that discriminate the organization from its competitors, qualifications of the key leaders, and description of the outputs from the organization. Its initial purpose was to attract funding for the organization.

As the organization gets established, there comes a growing need for it to develop a Strategic Business Plan to define what new products will be coming out over the next 5 to 10 years and what standard products will be put to bed. It also should define projected volumes/demands for the product or service. It usually is a five-year action plan that drives how the outside world will view the organization.

The more advanced organizations have realized that the organization's strategy needs to include both a five-year plan for how the outside world would view the organization and a five-year plan for internal operations that defines how the culture of the organization needs to change. The Combined PAM Plan provides the answer to this need. (In some organizations, the Combined PAM Plan is called the Strategic Improvement Plan.) The Combined PAM Plan is designed to bring

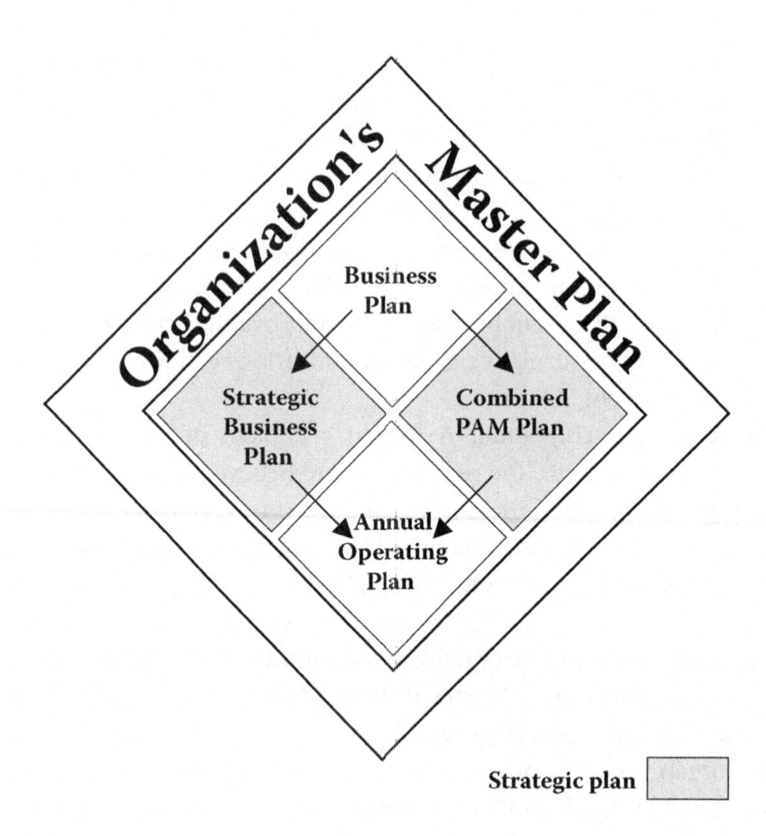

FIGURE 8.2
Organizational Master Plan.

about significant improvements in the KPDs of the organization. It is designed to establish more desirable behavioral patterns/habits at all levels of the organization.

When you blend together the Strategic Business Plan and the Combined PAM Plan, the organization has created a total Strategic Plan that will direct the evolution of the organization over the coming 5 to 10 years.

The fourth plan that exists in these advanced organizations is the Operating Plan. You will note we did not call it Annual Operating Plan. Most organizations do an Annual Operating Plan that is a bottoms-up analysis of the resources required to operate the organization over the next three years. The first year of the Annual Operating Plan is scrutinized in detail. At IBM, this activity started in July and, after many iterations of removing projects, adding projects, across-the-board cuts in spending, and many hours of presentation, eventually, just before the beginning of the new fiscal year in January, a firm Annual Operating Plan was signed off by the executive team and distributed throughout the organization.

Today many organizations have a Living Operating Plan. In these cases, the Operating Plan is altered frequently throughout the year based on shifts in schedules and customer demands. As projects are approved, dropped, or improvements are made, the approved adjusted resources are automatically included in the Operating Plan by Finance and a new Living Operating Plan is distributed. This eliminates the need for endless preparation of budget go-rounds by the line and middle managers. If the affected managers disagree with the new Operating Plan, they resolve the problems at the project level.

TWO MAJOR CONTINUOUS IMPROVEMENT METHODOLOGIES

Although continuous improvement is highlighted in Phase VII, the basic concept of "everyone trying to be the very best and continuously striving to be better" needs to penetrate the organization from the beginning of Phase I and live as a key value for the total life of the organization. Continuous improvement by itself should be resulting in 5 to 10% efficiency and effectiveness improvements throughout the total organization every year. Throughout Phase VI, many tools/methodologies were implemented. These tools/methodologies should have resulted in accelerating the rate that the organization improved. The gains resulting from the implementation of these tools is not an ending point, it is the starting point where the individuals affected by the tool need to continuously focus on their activities and the process to continuously improve. Often the tools/methodologies that make up the PAM plan is implemented by a group of individuals, but the continuous improvement process is a personal one—one for which each individual is responsible. Personal creativity and sense of accomplishment is a trademark of a world-class organization. Part of every individual's job description should be to use his or her creative mind to bring about performance improvement within the organization with specific emphasis on the tasks for which he or she is responsible.

We are often asked: "To what degree of creative participation should employees on the production line have?" Well, there are a number of companies whose production workers submit an average of two implemented improvement ideas per month. This is not just limited to the production line workers, but these organizations are able to get the same level of high participation in the support areas, such as Finance, Human Resources, Production Control, Executive Committee, Quality Assurance, etc.

There are many things that we could present related to stimulating continuous improvement within an organization, but, for purposes of this book, we will only discuss two of the major ones:

1. Area Activity Analysis
2. Organizational Alignment

Area Activity Analysis

> Everyone in IBM has customers, either inside or outside the company, who use the output of his or her job; only if each person strives for and achieves defect-free work can we reach our objectives of superior quality.
>
> **John R. Opel**
> *Past chairman of the board*
> *IBM*

Area Activity Analysis is probably the most effective and least used methodology of all the tools available to propagate continuous improvement and team operations. This is a methodology designed to align the authority, responsibilities, and accountabilities of each Natural Work Team within the organization. It is designed to establish effective internal and external supplier/customer relationships and operations.

Definitions

Area Activity Analysis (AAA): A proven approach used by each Natural Work Team (area) to establish efficiency and effectiveness measurement systems, performance standard improvement goals and feedback systems that are in line with the organization's objectives and understood by the employees involved.

Area: Any Natural Work Team that is organized to work together for an extended period of time. For example, it can be the organization's president and all the vice presidents and staff reporting to him or her. It also can be the maintenance foreman and all the maintenance workers reporting to him or her.

Natural Work Team (NWT): Sometimes called Natural Work Groups (NWG), is a group of people who are assigned to work together and report to the same manager or supervisor.

Most organizations talk about improving internal customer satisfaction, but do little to change the way they function. As soon as you ask people what they are actually doing to understand, plan for, and meet their internal customers' requirements, the conversation gets very quiet. Extending the concept of customer and supplier to the relationships within our organizations represents the next frontier of quality improvement. We are overdue to stop talking about the concept of internal customers and to start doing something about improving the internal supplier/customer relationships.

Usually, our employees have an excellent understanding of what they do and, as a result, they are normally the experts in their specific assignments. However, are they always doing the right thing on a day-to-day basis? Too often, employees waste resources because they do not understand how they fit into the total organization. They often view the organization as a complex puzzle that has never been put together.

AAA is the first performance improvement tool that a manager should use to help his or her area get started on a sound footing. AAA helps the organization accomplish the most basic of all management tasks. Defining AAA will help you to do the following:

- Clarify your NWT's real purpose.
- Identify those time-consuming activities that do and do not support your mission.
- Bring better alignment between your mission, activities, and the expectations of your internal and external customers.
- Align your employees' activities with the NWT's priorities.
- Identify which activities add real value and which can be minimized or eliminated.
- Understand how to make the transition from finding and fixing problems to preventing them.
- Clarify your requirements for your internal and external suppliers and measure their performance.
- Define a comprehensive measurement system for the critical activities that take place within your NWT and set performance standards for each of them.
- Put together an implementation plan to make it all happen.

Many of us are frustrated with our inability to accurately put a finger on the source of the problems we are facing every day. We know we work hard, deliver quality output, and drive the people who work for us to do the same. The harder we work and the more we push, the more entrenched the problem seems to get and employees' morale seems to sink. It is indeed a vicious, negative cycle that can sap the energies of even the most dedicated manager or employee. Here again AAA can help:

- AAA is a simple but powerful tool that you can begin using immediately to provide clear direction on what may otherwise have become a confusing journey toward improving customer service and aligns every unit with the organization's strategic plan.
- AAA can help you align your energy and resources with your organization's mission in a way that can result in greater effectiveness, efficiency, satisfaction, and teamwork.
- AAA serves as a compass to help you find your way through the jungle of overwork that threatens to overtake us all.
- AAA helps you sort out the vital few from the trivial many so that you can focus on delivering the value that you, and you alone, can add to your organization and your customers.
- AAA was designed by managers as a way to analyze and organize their work areas to get better results from their current resources. As the weeks, months, and years go by, the organizations we manage inevitably take on more and more responsibilities and our jobs get more and more complex. We begin to feel like we are running on a never-ending treadmill while someone keeps turning up the speed.
- AAA is unlike any other technique for improving processes, reducing costs, or decreasing turnover. It helps everyone clarify expectations and focus their efforts on the area's mission.
- AAA is an appropriate tool for new or existing areas or departments. It is a tool that will help to ensure that everyone understands their area's mission, customer expectations, what they need to do to succeed, and how to measure their performance.
- AAA can be used by managers at any level of the organization to improve the efficiency, effectiveness, and teamwork within their operations. It can be used by an individual unit or as part of a coordinated, organization-wide effort. It also can be used by an individual to improve his or her performance. It can and should be used by every person in the total organization at every level, from the team

of employees (vice presidents) who report directly to the president of the organization to the team of maintenance workers who report directly to the maintenance line manager.

- AAA is not another technique for improving processes, lowering cycle time, or reducing costs. Simply stated, it helps managers clarify what is expected of their groups, define the area's key measurements, set performance standards, and focus people's efforts like a laser beam on the organization's mission. It defines whether the area needs to improve and where the improvement opportunities exist. It helps the employees understand what is important for their customers and managers. It also helps the employees understand how they fit into the organization and contribute to the organization's goals. It is a people-building approach that helps them stand on their own feet with a high degree of confidence in themselves and helps them understand that they are doing something worthwhile.

Before we go any farther, let's discuss what we mean by customer/supplier relationships as they relate to internal and external customers. Basically, a customer/supplier relationship can develop in two different ways:

1. An individual or organization can determine that it needs something that it does not want to create itself. As a result, it looks for some other source (supplier) that will supply the item or service at a quality level, cost, and delivery schedule that represents value to the individual or organization (customer).
2. An individual or organization (supplier) develops an output that it believes will be of value to others. Then the individual or organization looks for customers that will consider the supplier's output as being valuable to them.

Two key points need to be made:

1. A customer/supplier relationship cannot exist unless the requirements of both parties are understood and agreed to. Too often, customers expect input from suppliers without understanding their requirements and/or capabilities without defining exactly what they need. On the other hand, too many suppliers provide output without

defining their potential customer's requirements and obtaining a common, agreed-to understanding of what both parties' requirements are.
2. Both the customer and the supplier have obligations to provide input to each other. The supplier is obligated to provide the item or service and define future performance improvements. The customer has an obligation to provide compensation to the supplier for its outputs and feedback on how well the outputs perform in the customer's environment.

The customer/supplier process has a domino effect. Usually, when a supplier is defined, that supplier requires input from other sources in order to generate the output for its customer. As a result, it becomes both a customer and a supplier (Figure 8.3).

Although the procedures related to internal customer/supplier partnerships are less stringent and have been simplified because the internal customer does not pay for the services that are provided, the concepts are equally valid. Too often, we set different standards for internal suppliers than we have for external suppliers. As a result, many of the internal suppliers provide outputs that are far less valuable than what it costs to produce the outputs. This often results in runaway costs and added bureaucracy. With AAA, we will show you how to apply the customer/supplier model to the internal organization, thereby improving quality and reducing cost and cycle time of the services and items delivered both within and outside the organization. It also serves to bring each area within the organization in line with the strategic plan.

Defining customer/supplier relationships is only one part of making an area function effectively. There are many other factors that also must be considered. For example:

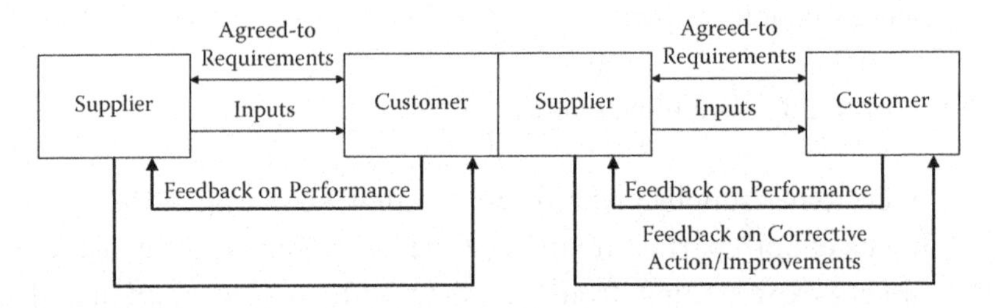

FIGURE 8.3
Feedback on corrective action/improvement.

- What is the area responsible for?
- How is the area measured?
- What is acceptable performance?
- How does the area fit into the total organization?
- How well do the area's employees understand their roles and the ways they can contribute?
- What are the important activities that the area performs from top management's standpoint?
- How does the area align itself with the organization's strategic plan?

It is important to note that all of these questions pivot around the activities in which the area is involved. It is for this reason that the AAA methodology broadened its perspective to go beyond the customer/supplier partnership concept to embrace a complete business view of the area.

The AAA methodology has been divided into seven different phases to make it simple for the NWT to implement the concept (Table 8.1). Each of these phases contains a set of steps that will progressively lead the NWT through the methodology.

> A people-building philosophy will make the program successful, a people-using philosophy will cause the program to fail.
>
> **Dr. Kaoru Ishikawa**
> *Creator of the cause and effect ("fishbone") diagram*

TABLE 8.1

The Seven Phases that Make Up the AAA Methodology

Phase	Number of Steps
Phase I: Preparation for AAA	5
Phase II: Develop Area Mission Statement	6
Phase III: Define Area Activities	8
Phase IV: Develop Customer Relationships	7
Phase V: Analyze the Activity's Efficiency	6
Phase VI: Develop Supplier Partnerships	5
Phase VII: Performance Improvement	8
Total	**45**

The Seven Phases of AAA

We will briefly describe each of the seven phases of AAA.

Phase I: Preparation for AAA

AAA is most effective when it precedes other initiatives, such as Organizational Alignment, Continuous Improvement, Team Problem Solving, Total Quality Management, Reengineering, or new IT systems. It is also best to implement the AAA methodology throughout the organization. This does not mean that it will not work if other improvement activities are underway or if it is only used by one area within the total organization. In the preparation phase, the good and bad considerations related to implementing AAA within an organization should be evaluated. If there is a major change in the organization's structure or to its major processes, AAA should be used. It also should be used in an organization that doesn't already have an excellent internal customer/supplier relationship. Based on the complexity and the percentage of people affected, a decision is made whether or not to use AAA within the organization. If the decision is made to use it, an implementation strategy is developed and approved by management.

Phase I is divided into five steps:

- Step 1: Analyze the Environment
- Step 2: Form an Implementation Team
- Step 3: Define the Implementation Process
- Step 4: Involve Upper Management
- Step 5: Communicate AAA Objectives

Phase II: Develop Area Mission Statement and Key Measurements

An organizational mission statement is used to document the reasons for the organization's or area's existence. It is usually prepared prior to the organization or area being formed, and is seldom changed. Normally, it is changed only when the organization or area decides to pursue new or different set of activities. For the AAA methodology, a mission statement is a short paragraph, no more than two or three sentences, that defines the area's role and its relationships with the rest of the organization and/or the external customer.

This activity starts at the very top and flows down through the organization like dominos falling over. It starts with the president comparing the mission statement for himself/herself and the people that report directly to the president to that of the organizations. The vice presidents then review each of their area's mission statements and compare them to the president's to be sure there is a common agreement on what the vice presidents' group was delegated to do. Next, the middle managers meet with the first-line managers and they review the middle managers' mission statements to be sure it supports the related vice presidents' mission statement and doesn't conflict or overlap with any of the middle managers' mission statements. Now that the middle managers' mission statements are approved, it is time for the first-level managers to meet with their people to prepare their mission statements. Again, their mission statements must support the middle managers' mission statements and not overlap with other first-level managers' mission statements. This is called the waterfall effect, which starts at the top and moves down through the organization (Figure 8.4).

Every area should have a mission statement that defines why it was created. It is used to provide the area manager and the area employees with guidance related to the activities on which the area should expend its resources. Standard good business practice calls for the area's mission

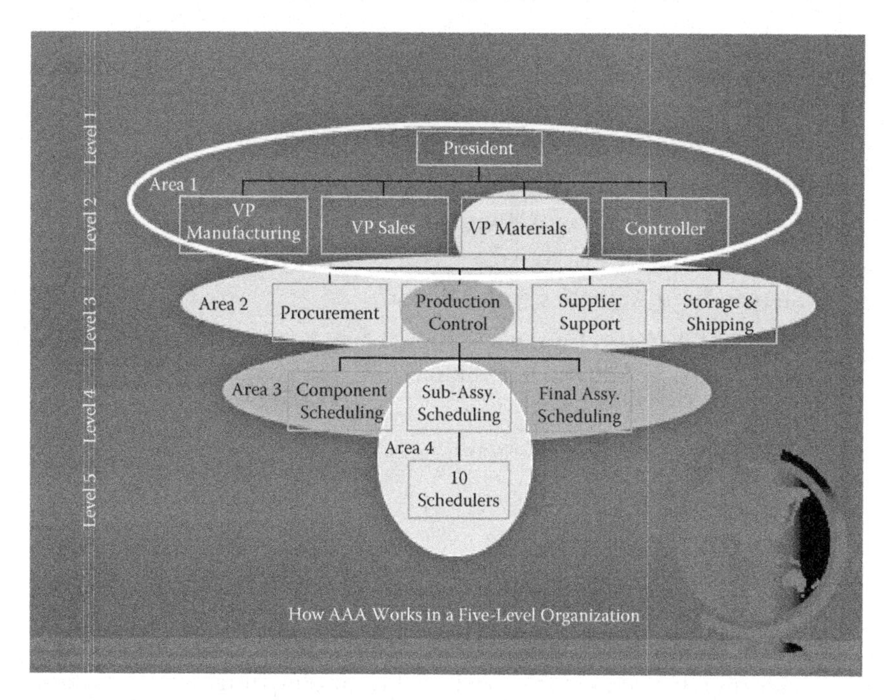

FIGURE 8.4
How AAA works in a five-level organization.

statement to be prepared before an area is formed. The mission statement should be reviewed each time there is a change to the organization's structure or a change to the area's responsibilities. It also should be reviewed about every four years, even if the organization's structure has remained unchanged, to be sure that the mission statement reflects the current activities that are performed within the area. If a mission statement does not exist, the AAA team will prepare a mission statement. In all cases, any change to the mission statement must be approved by upper management before it is finalized.

Also during Phase II, the area's service policy is developed. A service policy is a short statement that defines how the area will interface with its customers and supplier.

Phase III: Define Area Activities

During this phase, the AAA team will define the activities that are performed within the area. For each major activity, the AAA team will define the activity's output(s) and the customers that receive that output. Phase III: Define Area Activities is divided into eight steps:

- Step 1: Identify Major Activities (each individual in the area should do this)
- Step 2: Combine into Broad Activity Categories
- Step 3: Develop Percentage of Time Expended
- Step 4: Identify Major Activities
- Step 5: Compare List to Area Mission Statement
- Step 6: Align Activities with Mission
- Step 7: Approval of the Area's Mission Statement and Major Activities
- Step 8: Assign Activity Champions

Phase IV: Develop Customer Relationships

Definitions:

Supplier: An organization that provides a product (input) to the customer (Source: ISO 8402).

Internal Supplier: An area within an organizational structure that provides input into another area within the same organizational structure.

External Supplier: A supplier that is not part of the customer's organizational structure.

During this phase, the AAA team will meet with the customers that are receiving the outputs from the major activities conducted by the area to:

- Define the customer's requirements
- Define the supplier's requirements
- Develop how compliance to the requirements will be measured
- Define acceptable performance levels (performance standards)
- Define the customer feedback process

Phase IV is divided into seven steps:

- Step 1: Select Critical Activity
- Step 2: Identify Customer(s) for Each Output
- Step 3: Define Customer Requirements
- Step 4: Define Measurements
- Step 5: Review with Customer
- Step 6: Define Feedback Procedure
- Step 7: Reconcile Customer Requirements with Mission and Activities

Phase V: Analyze the Activity's Efficiency

For each major activity, the AAA team will define and understand the tasks that make up the activity. This is accomplished by analyzing each major activity for its value-added content. This can be accomplished by flowcharting the activity, and collecting efficiency information related to each task and the total activity. Typical information that would be collected includes:

- Cycle time
- Processing time
- Cost
- Rework rates
- Items processed per time period

Using this information, the AAA team will establish efficiency measurements and performance targets for each efficiency measurement. Phase V is divided into six steps:

- Step 1: Define Efficiency Measurements
- Step 2: Understand the Current Activity

- Step 3: Define Data Reporting Systems
- Step 4: Define Performance Requirements
- Step 5: Approve Performance Standards
- Step 6: Establish a Performance Board

Phase VI: Develop Supplier Partnerships

Using the flowcharts generated in Phase V, the AAA team identifies the supplier that provides input into the major activities.

This phase uses the same approach discussed in Phase IV, but turns the customer/supplier relationship around. In this phase, the area is told to view itself in the role of the customer. The organizations that are providing the inputs to the NWT are called internal or external suppliers. The area then meets with its suppliers to develop agreed-to requirements. As a result of these negotiations, a supplier specification is prepared that includes a measurement system, performance standard, and feedback system for each input. This completes the customer/supplier chain for the area.

Phase VI is divided into five steps:

- Step 1: Identify Supplier(s)
- Step 2: Define Requirements
- Step 3: Define Measurements and Performance Standards
- Step 4: Define Feedback Procedure
- Step 5: Obtain Supplier Agreement

Phase VII: Performance Improvement

This is the continuous improvement phase that should always come after an activity has been defined and the related measurements are put in place. It may be a full TQM (Total Quality Management) effort or just a redesign activity. It could be a minimum program of error correction and cost reduction or a full-blown Total Improvement Management project.

During Phase VII, the NWT will enter into the problem-solving and error-prevention mode of operation. The measurement system should now be used to set challenge improvement targets. The NWT should now be trained to solve problems and take advantage of improvement opportunities. The individual efficiency and effectiveness measurements will be combined into a single performance index for the area. Typically, the area's key measurement graphs will be posted and updated regularly. Figure 8.5 is a typical Performance Board for an individual area.

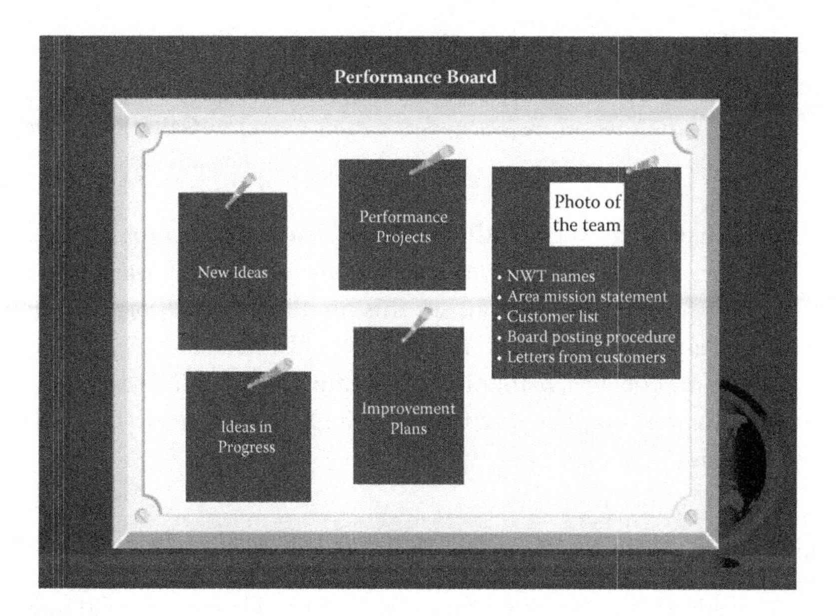

FIGURE 8.5
Typical Performance Board.

> If I was limited to only one tool or methodology to bring about massive change in an organization, that one tool would be Area Activity Analysis.
>
> **H. James Harrington**

Organization Alignment

Now, it is obvious that there will be a lot of change going on within organizations that are operating using an Organizational Master Plan. Even if your organization has advanced to the point that a Combined PAM Plan is being implemented, there will be many changes taking place within the organization and within the processes. These changes are bringing about many demands and stresses upon the present organization. To continuously maintain the momentum and sustain the gains that are made, it is often advisable to take a fresh look at the way the organization is organized. All too often organizations are reorganized based on the whims and desire for power of a few key managers within the organization. Often this type of approach creates havoc, delays progress, and is very expensive.

> We trained very hard, but it seemed that every time we are beginning to form up into teams, we would be reorganized. I was to learn later in life that we tend to meet many new situations by reorganizing. A wonderful

method it can be for creating the illusion of progress while producing confusion in effectiveness and demoralization.

Gaius Petronius
Roman author, 70 AD

Too often, Organizational Alignment is used to hide a problem and to provide the current management team with a scapegoat who they can blame the problems on. You can see this problem in small organizations all the way up to the election of the president of United States. To eliminate these types of problems, realigning the organization needs to be treated in a very systematic manner. The Organizational Alignment Cycle consists of six phases (Figure 8.6).

Phase I: Strategic Plan

The Strategic Plan includes the Strategic Business Plan and the Combined PAM Plan. This is a logical starting point as a Strategic Plan sets the direction for the organization for the next 5 to 10 years. The objective of Organizational Alignment is to optimize the organization's structure, processes, personnel, and recognition system to maximize the use of these resources, thereby producing the desired results and often even exceeding the goals.

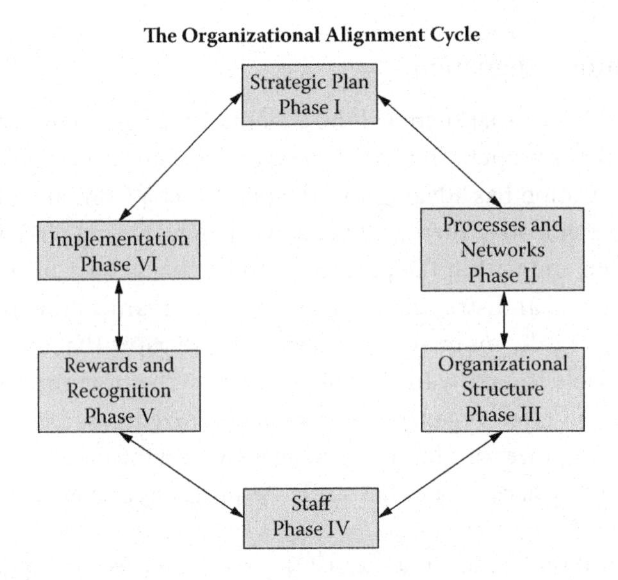

FIGURE 8.6
The six phases of the Organizational Alignment Cycle.

Phase II: Processes and Networks Design

In this phase, the way the organization functions and the circulation of information are developed and documented. The Strategic Plan defines where the organization needs to evolve to. The processes serve as the highway that the organization follows to get to the organization's goals. At this point in the Organizational Alignment Cycle, you need to look at these processes because they define how the organization operates. Each of the major processes needs to be examined to minimize the bureaucracy built into them and to eliminate the no-value-added activities. Processes are made up of activities that can be classified into three categories:

- Value-Added: These are activities that the customer would be willing to pay for.
- Business Value-Added: These are activities that are necessary to operate the organization, but are not directly related to what the external customer would be willing to pay for. (For example, processing employee payroll process, employee opinion surveys process, employee hiring process, accounts receivable processes, budgeting process, etc.)
- No-Value-Added: These are activities that add no value to the external customer or to the organization. (For example, rework, scrap, inspecting someone else's work, etc.)

The objective of the process reviews is to eliminate the no-value-added activities, minimize business value-added activities, reduce processing time and cycle time while improving the quality of the processes' output.

Organizations have recognized for many years the importance of focusing on improving their business processes, but in most cases have not realized the importance of optimizing the networks within the organization that provide communication and interfaces throughout the process. Flowcharting information flow and communication networks to optimize the way things get done are equally important and often overlooked.

Phase III: Organizational Structure Design

During this phase the team will redesign the organizational structure to determine the placement of power and authority within the organization. As process improvement teams work on streamlining the processes, it is often necessary to restructure the organization to minimize wasted effort and cycle time in moving items and information from one operating unit to another. Also, as the organization becomes more participative and the employees are empowered to act on their own, there is often a need to restructure the organization. In addition, a great deal of added benefits can be obtained by flattening the organization thereby greatly minimizing the distortion of information as it passes down the ladder.

There are 10 different types of organizational structures that are presently being used in different organizations. There is no correct one for all organizations. The 10 organization structures most commonly used today include:

1. Functional
2. Vertical
3. Bureaucratic
4. Decentralized
5. Product
6. Customer
7. Geographic
8. Case Management Network
9. Process-Based Network
10. Front-Back Hybrid

Each of these organizational options provides acceptable results depending on the culture, products, maturity, and complexity of the organization. Each of them has its own set of advantages and disadvantages. We have seen very successful companies that have worked with a mixture of these structures based on the needs of the different functions within the organization.

In designing a new organizational structure, we like to use a basic theme for the design. One of our favorite themes is Customer-Centric Chain (Figure 8.7).

For each of the five Cs in the Customer-Centric Chain there is a detailed process for getting the required information. For example, Figure 8.8 is the process for culture assessment.

Phase IV: Staffing Phase

This phase outlines the skills and mindsets required by the strategy and structure of the organization. Once the structure and its supporting processes have been defined and documented, it is important to understand

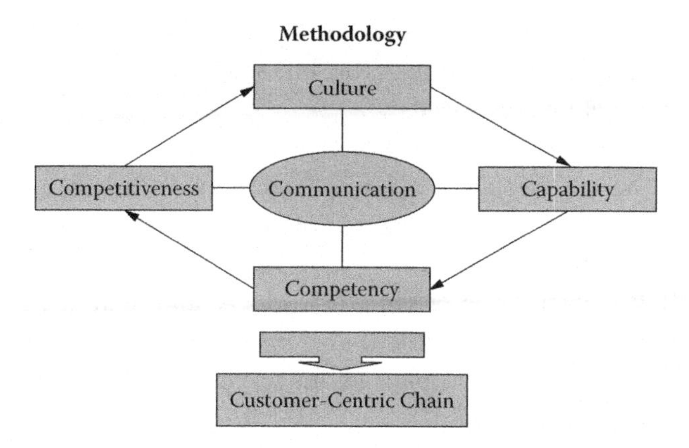

FIGURE 8.7
Customer-Centric Chain.

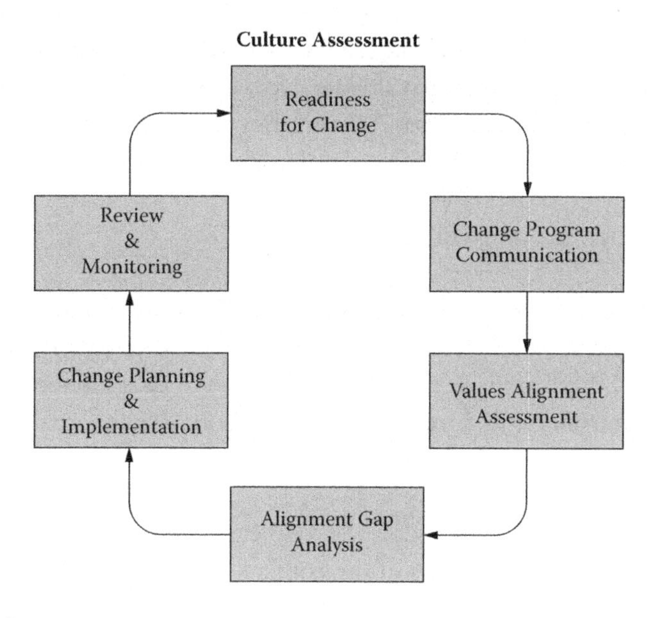

FIGURE 8.8
Process for culture assessment.

how well the people skills are aligned with the processes that they will be operating and using. To accomplish this, the organization should do a skills assessment of the present employees and then use the Strategic Plan to define the type of skills that will be needed in the future. Never mind what worked, or did not work in the past. The rules have changed. It is not enough to define how many accountants, salespeople, development engineers, or assembly workers you have versus the quantity that you will need five years from now. You need to define what skills they have or don't have that will be needed in the future. The following are typical questions that will help you define these needs:

- Will the salespeople need to have additional skills like experience accessing the Internet, creating PowerPoint® presentations, and conducting virtual conferences?
- Will we need fewer assembly workers, and will the ones who remained need to be highly skilled in computer operations?
- Will the marketing people and our service personnel need to speak Chinese and Russian?
- Will our purchasing personnel need to become our supply chain managers?

For sure, in most organizations, a much higher percentage of our people will need to become computer literate. Now is the time to start to plan for the needed skills change. As an organization, you have two options:

1. Train your personnel
2. Hire new people to fill the skills' void

The correct answer for most organizations is a combination of the two. First, stop hiring people who don't fill a skill's void, thereby letting attrition take care of the surpluses. If you need a person to do a job today that will result in a surplus skill, hire a temporary person or outsource it.

Phase V: Rewards and Recognition

This phase outlines the goals of the employee with the goals of the organization by providing motivation and incentives for the completion of the organization's strategic direction. The purpose of the reward and recognition system is to reinforce behavioral patterns that support the organization's mission, vision, and Strategic Plan. It is important to remember that the rewards and recognition system is designed to pay for performance, whether it is physical or mental, and to recognize people who have achieved something over and above what they are normally expected to accomplish. The individual's salary is a reward for a normal job well done. Only the truly outstanding people should receive contribution awards and special recognition.

Phase VI: Implementation

This phase defines how the changes are installed into the organization. It is the most critical phase of the total Organizational Alignment Cycle because all of the previous effort can go down the drain if the changes are not implemented effectively and accepted by the people who are affected by the changes. It is at this point that all the Organizational Change Management efforts you have been applying throughout the first five phases of the Organizational Alignment Cycle really pay big dividends.

> Only 5% of the organizations in the Americas and Europe truly excel. Their secret is not what they do, it's how they do it.
>
> **H. James Harrington**

Organizational Alignment is a basic business discipline that addresses both the processes and behaviors in a very systematic way that drives positive results. Its objective is to create an organization in which the employees and management devote their time and energy toward accomplishing results without wasting effort overcoming obstacles.

> Don't try to achieve complete alignment for that requires a very stable environment that is undesirable in most organizations.
>
> **H. James Harrington**

One organization that we have worked with that applied Organizational Alignment methods obtained the following results:

- Reduced two levels of management
- Reduced the management team size by 20%
- Improved customer satisfaction from 75 to 81%
- Established a Knowledge Management System
- Identified five processes that by outsourcing would reduce costs from 15 to 25% while obtaining higher-quality output
- Reduced cycle time and processing time on six processes

> The current work environment is rich in social, psychological, and political drivers that cause fatigue and eventually leads to burnout.
>
> **Barbara Kaufman**
> *President*
> *ROI Consulting Group*

PAM PROCESS SUMMARY

> I find that people spend more time and effort in planning and implementing their vacations than they do related to their work careers.
>
> **H. James Harrington**

For thousands of years, organizations have been searching for the magic bullet that will maximize the organization's performance. Literally thousands of books have been written documenting the best ways to accomplish this elusive task. Although this author is not familiar with all the books that have been written on performance improvement, the ones that he has read over the past 50 years have primarily been focused on tools and techniques that bring about short-range solutions, although many of them claim that they will bring about transformation in the organization's culture. To the best of the author's knowledge, the PAM process is the first approach that focuses on defining what the KPDs are within an organization and how these need to change over a period of years. Once the organization has defined what and how much they want to change, they can select the best tools, methodologies, and techniques to bring about the desired transformation to the organization's culture that is required to bring about the high levels of performance that the organization is trying to achieve. To date, the author has identified more than 1,300 different performance improvement approaches that have been used by various organizations within the past 50 years. All of them have proved to be successful under the right conditions at bringing about improvements in organizational performance. Many of these tools/methodologies are similar in the basic approach and only differ slightly on how they are implemented or the terms they use to describe the tool/methodology. Unfortunately, most organizations have chosen to follow the "flavor of the month" performance improvement approaches without really considering how they need to change the culture inside the organization to get the desired performance results. The major advantage that PAM process has over previous tools/methodologies is that it defines first what and how much the KPDs need to change over a period of time and then it looks at the 1,300+ tools/techniques to select the best combination to bring about the desired transformations at minimum cost and disruption to the organization. The PAM approach is designed to bring about short-term results, thereby strengthening the organization's commitment to the process as well as creating a major transformation to the organization's KPDs over a period of years. It, therefore, provides the best of both worlds: short-term results to meet management needs and long-term sustainability to meet the growth requirements of the organization.

> TQM failed, Six Sigma failed, and Lean is failing, but their basic principles endure forever as a way to bring about performance improvement. The fact that they are renamed and gain popularity is a testament to their legitimacy. Many people visit the Grand Canyon and are amazed by its grandeur, but never go back again because they already did that.
>
> **H. James Harrington**

Appendix A

DEFINITIONS AND ABBREVIATIONS

Annual Operating Plan: A formal statement of the organization's short-range goals, the reason they are believed to be attainable, the plans for reaching these goals, and the funds approved for each part of the organization (budget). The Annual Operating Plan is often just referred to as the Operating Plan.

Area: An area is any Natural Work Team that is organized to work together for an extended period of time. For example, it can be the organization's president and all the vice presidents and staff reporting to him or her. It also can be the maintenance foreman and all the maintenance workers reporting to him or her.

Area Activity Analysis (AAA): A proven approach used by each Natural Work Team (area) to establish efficiency and effectiveness measurement systems, performance standard improvement goals, and feedback systems that are in line with the organization's objectives and understood by the employees involved.

Business Plan (BP): A formal statement of a set of business goals, the reason they are believed to be obtainable, and the plan for reaching these goals. It also contains background information about the organization.

Combined PAM (Performance Acceleration Management) Plan: This plan focuses on how to change the culture of the organization. It is designed to answer the questions: How do we excel? How can we increase value to all stakeholders? It addresses how the organization's controllable factors (KPDs) can be changed to improve the organization's reputation and performance. Some organizations call this their Strategic Improvement Plan.

Controllable Factors (CF): A synonym for Key Performance Drivers (KPDs). They are the factors/things that management can invest in or remove assets from (money or effort) in order to bring about changes in the performance of the organization.

External Supplier: A supplier that is not part of the customer's organizational structure.

Focus Group: A group of people, who have a common experience or interest, is brought together where a discussion related to the item being analyzed takes place to define the group's opinion/suggestions related to the item being discussed.

Future Shock: The point at which no more change can be accommodated without the display of dysfunctional behaviors within the organization affected. It is a term frequently used in Organizational Change Management Methodology.

Internal Supplier: An area within an organizational structure that provides input into another area within the same organizational structure.

Key Performance Drivers (KPDs): Things within the organization that management can change that control or influence the organization's culture and the way the organization operates. (These are also called controllable factors.)

Mission Statement: The stated reason for the existence of the organization. It is usually prepared by the CEO and key members of the executive team. It is typically changed only when the organization decides to pursue a completely new market.

Natural Work Team (NWT): Sometimes called Natural Work Groups (NWG), is a group of people who are assigned to work together and report to the same manager or supervisor.

Organizational Master Plan: The combination and alignment of an organization's Business Plan, Strategic Business Plan, Combined PAM Plan, and Annual Operating Plan.

Personality: The impression or impact the individual or group has on other individuals or groups.

Roadblock: Something, such as a situation or condition, that prevents further progress toward an accomplishment.

Steering Committee: An abbreviation for the Performance Acceleration Management Steering Committee.

Strategic Business Plan: This plan focuses on what the organization is going to do to grow its market share. It is designed to answer the questions: What do we do? How can we beat the competition? It is directed at the product and/or services that the organization provides as viewed by the outside world.

Subcommittee: A team assigned to develop the Performance Acceleration Management (PAM) plan for each of the KPD's vision statements.

Supplier: An organization that provides a product (input) to the customer. (From: ISO 8402.)

Transition: An orderly progress from one state, condition, or action to another.

Value: The basic beliefs on which the organization is founded. The principles that make up the organization's culture are often called "values." These are prepared by top management. They are rarely changed because they must be statements that the stakeholders hold and depend on as being sacred to the organization.

Vision Statement: A vision statement is a documented view of the future desired state condition. A vision statement should be a short paragraph no more than three sentences long. (A one-sentence vision statement is preferable.) A vision statement should stretch the organization to be the best that it can be. Documented vision statements provide an effective tool to help develop objectives and improvement programs.

Appendix B

PARTIAL LIST OF OVER 1,400 DIFFERENT PERFORMANCE IMPROVEMENT TOOLS AND METHODOLOGIES

This appendix lists more than 1,474 Performance Improvement approaches.

5 S
5 Whys (5Ws)
5 Ws and 2 Hs Method
6-3-5 Method
7-S Model
80/20 Rule
A Delta T
Accelerated Solution Environment (ASE)
Accelerated System Development
Acceptance Decisions
Acceptance Sampling
Accountability Matrix
ACORN Test
Acquisition Streamlining
Action-and-Effect Diagram
Action Diagramming
Action Planning
Action Readiness Chart
Activity Accounting
Activity Analysis
Activity-Based Budgeting
Activity-Based Costing
Activity-Based Management
Activity Chart
Activity Cost Matrix
Activity Cost Pool Definitions
Activity Network Diagram
Activity-On Arrow
Activity-On Node
Actual Cost of Work Performed
Add-On/Replacement Matrix
Affect-Task Concept Balancing
Affinity Analysis
Affinity Diagrams
Algorithms
Alignment Processes
Amoeba Units
Analogy and Metaphor
Analysis and Segmentation of Customer Views
Analysis of Customer Wants
Analysis of Variance
Annual Strategic Quality Plans
ANOVA
Application Construction
Application Context Diagram
Application Development Alternative

Application Evolution
Application Installation
Application Prototype
Application Screen Flow
Application Structuring and Identification
Application Testing
Architecture for Managing Corporate Culture
Architecture Interaction Diagram
Area Activity Analysis (AAA)
Area of Impact
Arrow Diagrams
Assess Change Management Enablers and Barriers
Assets Management
Assimilation Capacity Audit Procedures
Assimilation Capacity Audit Software
Assimilation Capacity Consultation
Association Diagram
Association Matrix
Association Programming
Assumptions Evaluation
Attribute Acceptance Sampling
Attribute Control Charts
Attribute Identification
Attribute Measurement
Attribute Sampling
Attribute Sampling Procedures
Attribute Sampling Tables
Audience Analysis
Auditing
Audits by Top Management
Automatic Test Equipment
Automation
Autonomous Work Teams
Auxiliary
Average
Average (x) Chart
Awards
Axiomatic Design
Balance Sheet
Balanced Score Card (BSC)
Baldrige, Award
Bar Chart
Barplots
Barrier-and-Aids Analysis
Baseline Load Factor
Basic Quality Functions
Basili Data Collection Analysis
Batch Procedure Design
Bayes' Theorem
Bayesian Estimates
Behavior Model
Behavior Modification
Behavior of a System
Behavior Pattern
Behavioral Analysis
Benchmarking
Benefits Assessment
Bessel Function
Best-Value-Future-State Solution
Big-Picture Analysis
Binomial Distribution
Bivariate Distribution
Black Belt Training
Black Box Testing
Block Diagram
Block (random) Sampling
Blue-Collarization
Bottom-Up Testing
Box Plot
BPI Measurement Methods

Brainstorming
Brainwriting Pool
Breakdown Tree
Budget Attainment Analysis
Bureaucracy Elimination Methods
Business Area Data Modeling
Business Case Development
Business Drivers
Business Engineering
Business Performance Management (BPM)
Business Performance Management Measurement
Business Plan
Business Process Improvement Concepts
Business Process Improvement Measurement Methods
Business Process Innovation
Business Process Reengineering
Business Process Simulation
Business Strategy Analysis
Business Systems Planning (BSP)
Business to Business (B2B)
Business to Customer (B2C)
Business to Employee (B2E)
Business Transaction Identification
Buzz Group
c Charts
Cp
Cpk
Calibration
Canonical Synthesis
Capacity and Staff Planning
Career Development
Career Planning
Cascade Processes
CASE (Computer-Aided Software Engineering)
Case Study
Cash Bonuses
Catastrophe Effects Diagram
Causal-Loop Analysis
Causal-Loop Diagram
Cause/Effect Graphing
Cause-and-Effect Diagrams (Fishbone Diagrams)
Cause-and-Effect Diagrams with Cards (CEDAC)
Cause-and-Effect Matrix
Central Composite Designs
Central Limit Theorem
Central Tendency Measurement
Chain Sampling Plans
Change Agent Evaluation
Change Agent Selection Form
Change History Survey
Change Impact Evaluation
Change Implementation Monitoring
Change Knowledge Assessment
Change Leader Readiness Assessment (CLR)
Change Management Architecture Consultation
Change Program Portfolio (CPP)
Change Project Analysis
Change Project Evaluation Service
Change Resistance Scale
Change Synergy Evaluation
Charts
Check List
Checkerboard Method

Checksheet Design
Checksheets
Chi-Square Distribution
Chi-Square Distribution/Test
Circle of Influence
Circle of Knowledge
Circle of Opportunity
Circle Response
Class Relationship Diagram
Classification of Characteristics
Clearance Fits
Clearing Interval
Cluster Analysis
Cluster Factor Assessment
Cluster Organizations
Coaching
Coaching Styles Inventory
Code Generation
Coefficient of Contingency©
Cognitive Architecture
Cognitive Health
Cognitive Quality-of-Worklife
Cognitive Technologies
Cognitively Balanced Jobs and Careers
Collaboration Technology
Collecting Data
Communicating Change
Communicating Styles Survey
Communication
Communication Management and Planning
Communication Techniques
Communications Planning
Company-Based Training
Company-Wide Quality Control
Comparative Analysis
Comparison Matrix
Competency Analysis
Competency Gap Assessment
Competency Model
Competitive Analysis
Competitor Product Disassembly Research
Computers
Computer-Aided Design (CAD)
Computer-Aided Engineering (CAE)
Computer Augmentation of Cooperative Work
Computer Forms of Cooperative Work
Computer Model
Computer Simulation
Computer-Supported Cooperative Work
Concept Diagram
Concept Ontology
Conceptual Data Model
Conceptual Process Partitioning Model
Concurrent Engineering
Conditional Probability
Conditions-of-Doing Effects
Conference Room Test
Confidence Intervals
Configuration Management
Conflict Management Guidelines
Conflict Resolution Diagram
Conjoint Analysis
Connectivity Analysis
Consensus
Consensus Building
Consensus Decision Making
Consensus Design
Conservation Analysis
Constant

Constituency Analysis
Consultation Map
Consultation Net Analysis
Consumer Risk Quality (CRQ)
Context Diagramming
Contextual Leadership
Contingency Diagram
Contingency Planning
Contingency Reserve
Contingency Tables
Continuous Flow Manufacturing
Continuous Sampling Plan
Continuum of Team Goals
Contract Negotiation
Control Array
Control Charts
Control Limits
Control of Quality
Controllable Factors
Conventional True Value (of a Quantity)
Coordination Technology
Corporate Governance
Corrective Action
Corrective Evolution
Correlation Analysis
Correlation Coefficient
Cost/Benefit Analysis
Cost-Cycle Time Analysis
Cost-Driven Analysis
Cost-Effectiveness Programs
Cost Flow Diagramming
Cost of Conformity
Cost of Lost Opportunity
Cost of Low Standards
Cost of Nonconformances
Cost of Nonconformity
Cost of Poor Quality
Cost-of-Quality Analysis
Cost, Quality, Features, and Availability
Cost-Time Analysis
Cost-Time Charts
Could Cost
Countermeasures Matrix
Counting Rules
Crawford Slip Method
Creative Brainstorming
Creative Thinking
Creativity Assessment
Criteria Filtering
Critical Dialogue
Critical Incident
Critical Path Method (CPM)
Critical Success Factor Analysis
Critical-to-Quality Analysis
Criticality Analysis
Cross-Functional Hobbying
Cross-Functional Management
Crossplots
Cross Unit
Cross-Unit Promotion Paths
CRUD Matrix
CSA/SD
Culture Alignment Assessment
Culture Assessment
Culture Audits
Culture Consistency Evaluation
Culture Roadblocks
Cumulative Distribution Function
Cumulative Hazard Sheet
Cumulative Sum Control Chart
Current Reality Trees
Current State Analysis
Current State Mapping
Current Systems Investigation

Customer Acquisition/ Defection Matrix
Customer Analysis
Customer Data Analysis and Action Plans
Customer/Employee Surveys
Customer-First Questions
Customer Interface Training
Customer Leadership Establishment Chart
Customer Loss Analysis (Cost Impact)
Customer Needs Table
Customer Partnerships
Customer Phone Calls (Management and Employees)
Customer-Related Measurements
Customer Relationship Management (CRM)
Customer Requirements Mapping
Customer Reviews
Customer Round Tables
Customer Satisfaction Analysis
Customer Satisfaction Ratings
Customer Satisfaction Standards
Customer Simulated Testing
Customer Surveys
Customer Understanding Tour
Customer Visits
CUSUM Chart
CUSUM Control Charts
Cycle Time Analysis and Reduction
Cycle Time Flowchart
Cycle Time Reduction Methods
Cycle Time Reporting
DBMS (Oracle, etc.)
Daily Life Protocol
Data Access Modeling
Data Analysis
Data Collection
Data Collection Strategy
Data Flow Diagramming
Data Gathering by Document Review
Data Gathering by Interview
Data Gathering by Samples and Surveys
Data Gathering by Secondary Research
Data Sheets
Data Stratification
Data Structure Programming
Data Structuring Diagram
Data Usage Matrix
Data View Identification
Data Wall
Database Generation
Decade Colleges
Decision Limits
Decision Process Flowchart
Decision Table
Decision Tree Diagram
Defect Map
Define, Measure, Analyze, Design, Verify (DMADV)
Define, Measure, Analyze, Improve, Control (DMAIC)
DELAY1 Function
DELAY3 Function
Delphi Methods
Delphi Narrowing Techniques
Deming 14 Points
Democratic Rules of Order

Demographic Analysis
Dendogram
Department Improvement Teams (DIT)
Deployment Chart (Down-Across)
Descriptive Statistics
Design for Maintainability and Availability
Design for Manufacturing (DFM)/Design for Assembly (DFA)
Design for Six Sigma (DFSS)
Design for X (DFX)
Design Mapping
Design of Experiments
Design Quality
Design Review
Design to Production Transition
Detachment Costing
Determination Test
Development Group Hobbying
Different Point of View
Differing Viewpoints on Change: A MOC Data Byte
Dimensions Cube
Discipline Stretch Exercises
Discrete Event Simulation
Discrete Probability Distributions
Discrete Uniform
Dispersion Measures
Distribution Analysis
Distributions
DLINF3 Function
Documentation Design
Dodge–Romig System
Doing What You Say Discipline
Dot Diagram
Double Reversal
Double Sampling
Double Sampling Plans
Double Specification Limit Plans
Doubling Time
Duration Compression
Dynamic Behavior
Dynamic Structures
DYNAMO
E-Commerce
Earned Value Analysis
Economics of Quality
Education Cost Sharing
Education Design and Development
Effective-Achievable Matrix
Effective Delegation
Effective Listening
Effort/Impact Analysis
Electronic Facilitation of Meetings
Empirical Quantile Plots
Employee Empowerment
Employee (Satisfaction) Surveys
Employee Surveys
Employee Training
Employee Visits
Empowerment Path Matrix
End-User Visual Programming-by-Assembly
Endurance Matrix
Engagement Costing
Enterprise Application Integration Solutions (EAI)
Enterprise Backbone Network
Enterprise Data Model
Enterprise Information Model
Enterprise Operations Model

Enterprise Organization Model
Enterprise Resource Planning (ERP)
Enterprise Strategy Assessment
Enterprise Strategy Model
Enterprise Strategy Planning
Enterprise Value Model
Entity Analysis
Entity Life Cycle Analysis
Entity Life Cycle Diagram
Entity Relationship Diagram
Entity State Transition Diagram
Environmental Aspect
Environmental Controls
Environmental Impact
Environmental Management System Audit
Environmental Pride Process (Your Living Rooms)
Environmental Stress
Equilibrium
Equilibrium Value
Equipment Certification
Error Guessing
Error-Proofing
Error-Proofing Fixtures and Methods
Established Reliability
Estimation
European Quality Award
Event-On Node
Events Log
EVOP Evolutionary Operations
Executable Module
Executive Error-Rate Reduction
Executive Improvement Teams (EIT)
Executive Information Needs Analysis
Executive Needs Analysis
Exemplary Facilities
Expanded Uncertainty
Expected Values
Expert Decision Systems and Knowledge Bases
Expert System Voice Determination Matrix
Exponential Decay or Growth
Exponential Distribution
Exponential Formula for Reliability
Exponentially Weighted Moving Average (EWMA)
EXPRESS
External and Internal Customers
F Distribution
Facilitation of Teams
Facilitative Management
Facilitator Training
Facilitators
Facilities Planning
Facility Layout Diagram
Factor Analysis
Factorial Design
Factorial Experiment (General)
Fail-Safe Planning
Failure Analysis
Failure Mode and Effect Analysis (FMEA)
Failure Prediction
FAST (Fast Action Solution Teams)
Fault Diagnosis
Fault Modes, Effects, and Criticality Analysis (FMECA)
Fault Tree Analysis
Fewer Good Suppliers

Field Reporting
Financial Analysis
Financial Reporting
Finish-No-Earlier-Than Constraint
Finish-to-Finish Dependency
Finish-to-Start Dependency
Finite Element Analysis (FEA)
First-Order Delay
First-Time Yield (FTY) or Rolled Through Yield (RTY)
Fishbowls
Flattening the Organization
Flowcharting
Flow Diagram
FOCUS
Focus Groups
Focused Process Improvement
Focused Restructuring
Fog Index
Foolproof Engineering Methods
Force-Field Analysis
Forced Association
Forced Choice
Forms Design
Foundation Immersion Reinforcement
Fractional Acceptance Numbers Plan
Fractional Factorial Experiments
Frequency Distribution
Fresh Eye
Full Factorial Designs
Function Diagrams
Function Execution by Mass Work Events
Functional Map
Fundamental Concepts
Fundamental Process Sequence
Fury Logic
Future Reality Diagram
Future State Business Process Model
Future State Mapping
Future State Vision
Futures Wheel
Gain Sharing
Gantt Chart
Gap Analysis
Gaussian Curves
General Ledger Analysis and Consolidation
General Surveys
Geometric Dimensioning and Tolerancing
Geometric Distributions
Geometric Sample/Directional Sample
Global Model
Goal Control Process
Goal Gap
Good Manufacturing Practices (GMPs)
Goodness of Fit of a Distribution
Gozinto chart
Grace Latin Square
Graeco-Latin Designs
Graphical Evaluation and Review Technique (GERT)
Graphical Methods
Graphs
Group Doctorate
Group Interfaces
Group Presentation
Group Recognition
Group Sequential Plans

Groupware
Harrington's Ten Fundamental Tools
Harrington's Ten Sophisticated Tools
Hawthorne Effect
High-Impact Teams (HIT)
Histograms
Historical Analysis
History of Quality
HIT (High Impact Teams)
Horizontal Careers
Hoshin Planning
House of Quality
Human Empowerment Analysis
Human Factors Engineering
Hypergeometric
Hypergeometric Distribution
Hypertext
Hypothesis Testing
Idea Advocate
Idea Borrowing
Idea Grid
Idea Rooms
Implementation Architecture Program
Implementation Factors for Success Assessment (FSA)
Implementation History Assessment (IHA)
Implementation Plan Evaluation
Importance-Performance Matrix
Importance Weighing
Improvement Portfolio Development
Improvement Process Measurements
Improvisation Map
Indifference Quality Level
Individual Performance Indicators
Individual Public Recognition
Individuals (x) Chart
Industrial Combinatorics
Industrial Modernization Incentives Program (IMIP)
Industry Best Practices
Inference Testing
Influence Diagram
Influence Style Survey
Information Engineering
Information Model View
Information Needs Analysis
Information System Diagram
Information Warehouse Data
Innovation Training
In-Process Data Collection
In-Process Inspection
Inspection and Testing
Inspection Planning
Inspiring
Instructional Design Model
Instruments
Integrated Cost/Schedule Reporting
Interference Fits
Internal Customer Concept
Internal Customer Measurements
Internationally Harmonized Standards
Interrelationship Diagraphs (ID)
Interviewing Techniques
Intrinsic Error (of a measuring instrument)
Inventory Turnover Rate

Investment Management
Is-Is Not Matrix
Ishikawa Diagram
ISO 14000 Environmental Management Systems
ISO-9000 Compliance—ISO-2959-1; ISO-3534-2; ISO-3951; ISO-8402; ISO-19011
IT Applications
I-TRIZ
Iterator/Recursor Analysis
Job Descriptions
Job Training and Certification
Job-Related Training
JUSE
Just-In-Time (JIT)
Kaizen
Kaizen Blitz
Kanban
Kano Method; Total Quality Creation
Kano Model
Kendall Coefficient of Concordance
Key Performance Indicator (KPI)
Knowledge-Based Information Model
Knowledge Fractality Chart
Knowledge Management
Kolmogorov-Smirnov Test for Normality
Kruskal-Wallis One-Way Analysis
Kume Quality Rules
Laboratory Accreditation
Landscape Survey
Latin Square Model
Leader Alignment Planning
Leadership Effectiveness
Leadership Skills Development
Leadership Skills Training
Lean
Lean Management
Lean Metric
Lean Performance Indicator
Lean Six Sigma
Lean Thinking
Learning-Enablement Management
Levene Test
Life-Cycle Assessment
Life-Cycle Costing
Life-Cycle Costs/Costs of Ownership
Life-Cycle Impact Assessment
Life-Cycle Inventory Analysis
Limiting Quality; Limiting Quality Level
Line Chart
Line Management
Linear Relationship
Linear Responsibility Chart
Linking Diagram
List Reduction
Listening
Location Plots
Logic Diagram
Logical Database Design
Logical Relationship
Lognormal Distribution
Long-Range Quality Planning
Loss Function
Loss Function; Lot Tolerance Percent Defective (LTPD)
Lot Plot
M-Business
MRP-2

MX Bar–MR Charts
Maintainability
Maintainability Analysis
Maintainability Assessment
Major Process and Infrastructure Alignment
Major Program Status
Management by Objectives
Management by Objectives Achievement Measurement
Management by Walking Around
Management Improvement Teams (MIT)
Management of Change
Management Presentations
Management Process Alignment
Management Review and Approval
Management Self Audits
Management Theory History
Management's Seven Tools
Manufacturability Assessment
Manufacturing Resource Planning II
Mann-Whitney U Test
Market Analysis
Market Segmentation
Markov Analysis
Mastery Timeline
Matched Plans, derivation of
Matrix Charts–Decision Matrices
Matrix Data Analysis
Matrix Diagrams
Matrix Management
Maturity Grid:
Customer Interface
HR
Innovation
IT
Knowledge Management
Management
Measurement
Processes
Sales
Supplier Management
Training
Maximum Standard Deviation
Mean Time Between Failures (MTBF)
Measure of Dispersion
Measurement; central tendency, robust design
Measurement Error
Measurement Systems Analysis (MSA)
Measurement Matrix
Measurement Tools
Median Charts
Median Range
Median x Chart
Meeting Management
Mental Imaging
Mentorship Coverage Map
Meta-Action Diagram
Meta-Modesty
Method of Least Squares
Metrics Analysis
Metrology
Microdiagnostics
Middle-Up-Down Management
Milestones Chart
MIL-Q-9858A Compliance
MIL-HDBK-217,338; MIL-STD-105D,105E,756,756B,781,785,

1472, 1543, 1629/35 Compliance
Mind Flow
Mind Maps
Mission Statement
Mission Statement Checklist
Mission Statement Wordsmithing
Mistake Proofing
Mixture Designs
Mobilization Processes
Model (Organizational Effectiveness)
Model Building
Modeling
Modified Acceptance Control Charts
Modified Control Limits
Module Structure Diagram
Monastic Management
Monetary Awards
Monte Carlo Analysis
Monte Carlo Sampling
Monthly Assessment Schedule
Mood's Median Test
Morphological Analysis
Morphological Forecasting
Motivating the Work Force
Moving Average
Moving Range
Multilevel Continuous Sampling
Multinomial
Multiple-Attribute Decision Modeling
Multiple Correlation
Multiple Linear Regression
Multiple Rating Matrix
Multiple Regression
Multiple Sampling
Multiple Sampling Plan
Multiskills Maintenance
Multiskills Operator
Multistage Cluster Sampling
Multistage Sampling/Nested Sampling
Multivari Analysis
Multivari Strategy
Multivariable Analysis
Multivariable Chart
Multivariate Analysis
Multivariate Frequency Distribution
Multivoting
Mutually Exclusive Relationships
Mystery Shoppers
Narrow-Limit Gauging
National Association of Testing Authorities (NATA)
Needs Analysis
Needs-Seed
Negative Analysis
Negative Binomial
Negative Loop
Negotiation Techniques
Network Analysis
Neural Net Emotion Detection
Neural Nets and Classifier Search Engines
New Employee Selection
New Performance Standards
Noise Plan; Noise variables
Nominal Group Techniques
Nominal Prioritization
Nomograph Continuous Sampling Plans (CSP)

Nonconformances; Nonconforming Spacing
Nondestructive Testing
Nonlinear Relationships
Nonparametric Tests
Nonvalue-Added Activity and Bureaucracy Elimination Methods
Nonverbal Communications
Normal Distribution
Normal Probability Distribution
Normal Probability Plots
Notched Box Plots
No-Value-Added Analysis
np charts
Null Hypothesis
Numerical Prioritization
Object Behavior Scenario
Object Interaction Diagram
Objectives Matrix
Observation
One-Piece Flow
Online Computer-Aided Training
Online Conversion Design
Ontological Programming
Operating Characteristic Curve
Operating Efficiency
Operation Verification
Operational Alternatives Analysis
Operational Definitions
Opportunity Analysis
Opportunity Cycle
Order of Magnitude Estimate
Organization Chart
Organization Design
Organization Design and Structure
Organization Fractality
Organization Mapping
Organization Model
Organization Readiness Chart
Organization Relationship Map
Organization Taguchi
Organizational Alignment
Organizational Analysis
Organizational Breakdown Structure (OBS)
Organizational CAD
Organizational Change Enablers
Organizational Change Management (OCM)
Organizational Change Management Evaluation Process
Organizational Change Management Process Model
Organizational Change Management Techniques
Organizational Culture Diagnosis
Organizational Excellence
Organizational Human Resource Enablement Sessions
Organizational Improvisation
Organizational Knowledge Matrix
Organizational Neurosis
Organizational Performance Lever
Organizational Performance Lever (OPL) Dashboard
Organizational Prototyping
Organization's Master Plan
Orthogonal Arrays
Orthogonal Polynomial

Other Points of View (OPV)
Outsourcing
Overall Equipment Effectiveness
Overload Index
P Charts
Package Application
Package Evaluation
Package Integration Strategy
Package Software Evaluation
Package Validation Testing
Pain Management Strategies
Pain-Sharing Matrix
Pair Matching Overlay
Paired Comparison
Panel Debate
Paperwork Simplification Techniques
Paradigm
Paradox Analysis
Paradoxon in Smallest Work Unit
Parametric Estimating
Pareto Diagrams
Pareto Principle
Participative Management
Participatory Cabaret
Participatory Research Assembly
Participatory Town Meeting
Pattern and Trend Analysis
Pay for Knowledge System
Pay for Performance
Pearson Coefficient of Correlation
Peer Evaluations
People Enabler Detailed Analysis
Perfection Evolution
Performance Improvement Plan (PIP)
Performance Index
Performance Planning and Evaluation
Performance Qualification
Performance Standards (PS)
Periodic Systematic Sample
Personal Power Survey
Personal Resilience Questionnaires
Personality Profile (Kersey-Bates)
PERT Charting
Pessimistic Time
Phillips 66
Physical Database Design
PIC-A-Solution
Pictograph
Pie Chart
Pin Cards Technique
Plackett–Burman Designs
Plan-Analyze-Streamline-Implement Continuous Improvement (PASIC)
Plan-Do-Check-Act Cycle
Plan-Results Matrix
Planned Experimentation
Plus Minus Interesting
Point and Interval Estimation
Point-Scoring Evaluation
Poisson Distribution
Poisson Series
Poka Yoke
Policy Deployment
Polygon
Polygon Overlay
Poor-Quality Costs (PQC)
Portfolio Project Management

Positive Loop
Potential Problem Analysis (PPA)
Practical Intelligence Matrix
PRE–Control
Precontrol Techniques
Precedence Diagramming Method (PDM)
Prediction Intervals
Predried Sample
Prerequisite Tree
Presentation
Principal Component
Principal Component Analysis
Prioritization Matrices
Prioritization Through Ratings
Probability Concepts
Probability Density Function
Probability Distributions
Probability Plots
Problem Analysis
Problem Selection Matrix
Problem Solving
Problem-Solving Unit
Problem Specification
Problem-Tracking Logs
Problematic Behavior
Process Action Diagram
Process Analysis and Improvement
Process Analysis Technique (PAT)
Process Benchmarking
Process Capability Analysis
Process Capability Studies
Process Control Techniques
Process Decision Program Chart
Process Deployment Automation
Process Design Program Charts
Process Documentation
Process Elements
Process Empowerment Rooms
Process Engineering
Process Evaluation
Process Failpoints Matrix
Process Flow Controls
Process Flow Diagram
Process Improvement Teams (PIT)
Process Mapping
Process Modeling
Process Performance Matrix
Process Qualification
Process Redesign
Process Reengineering
Process Route Table
Process Selection Matrix
Process Simplification Techniques
Process Visioning
Process Walk-Through Methods
Process Window Definitions
Producers Risk Quality (PRQ)
Product Cycle Controls
Product Design Assurance
Product Liability
Product Metamorphic Transposition Matrix
Product Quality, eight dimensions of
Productivity Processes
Proficiency Testing
Program Evaluation and Review Technique (PERT) Charting

Program Group Dependency Diagram
Project Charter
Project Decision Analysis
Project Financial Benefits Analysis
Project Management
Project Management Knowledge Base
Project Plan
Project Planning Log
Project Prioritization Matrix
Project Role Map
Project Selection Matrix
Projection Analysis
Prototype Test Checklists
Prototyping
Pugh Concept Technique
Purposing Matrix
Qualitative Analysis
Qualitative Factor
Qualitization of Systems
Quality
Quality Area Improvement
Quality Assessment
Quality Assurance Manual
Quality Assurance Planning
Quality Awards
Quality Characteristics
Quality Chart
Quality Circle
Quality College
Quality Communication
Quality Company Policies
Quality Control
Quality Control Circles (QCC)
Quality Data Collection and Reporting System
Quality Engineering Methods and Training
Quality Function Deployment
Quality Improvement
Quality Improvement Teams (QITs)
Quality Integration
Quality Loss Function
Quality Management System (ISO 9000)
Quality Manuals
Quality of Management
Quality of Service
Quality Policy
Quality Policy Deployment
Quality Principle
Quality Related Costs
Quality Spiral
Quality Surveillance
Quality Survey
Quality Systems
Quality Visions
Quantile Plots
Questionnaires
Quick and Easy Kaizen
Quick Changeover (Single Minute Exchange of Die)
Quincunx; Quincunx Data
R Charts
Radar Chart
Random (Block) Sampling
Randomized Block Plans
Randomness Testing
Random Numbers Generator
Range Chart
Ranking Matrix
Rating Matrix
Rational Subgroups
Recall

Reduced Inspection
Redundancy Techniques
Reengineering
Regression Analysis
Regression Testing
Regularized Structure Reading Diagram
Reject Control Charts
Rejectable Process Level (RPL)
Relations Diagram
Relationship Map
Reliability Analysis
Reliability Block Diagram
Reliability-Centered Maintenance
Reliability Management System
Reliability Model
Reliability Predictions
Remote Maintenance/Telemaintenance/Online Maintenance
Report Design
Reproducibility Standard Deviation
Request for Corrective Action (RCA)
Requirements
Requirements-and-Measure Tree
Requirements Matrix
Requirements of Society
Research Assembly
Resiliency Management
Resource and Activity Driver Analysis
Resource Histogram
Resource Life Cycle
Resource Requirements Matrix
Response Data-Encoding Form
Response Matrix Analysis
Response Surface Design
Response Surface Methodology (RSM)
Responsibility Assignment Matrix
Responsibility Charting
Responsibility Matrix
Restructuring of the Quality Assurance Organization
Retainage
Reverse Brainstorming
Reverse Engineering
Reverse Thinking
Rewards and Recognition
Ringi Elimination
Risk Analysis
Risk Assessment
Risk Control
Risk Management Plan
Risk/Opportunity Management Process
Risk Space Analysis
Risk-Taking
Robust Design Approach
Room Collection Theory of Organizations
Root-Cause Analysis
Rotating Roles
Rotation Patterns
Round Robin Brainstorming
Rules of Probability, Combinatorics
Run Chart
Run-It-By
Rural Industriality
Safe-Life Concept
Safety Management Systems
Sample

Sample Averages/Range
Sample Inspection
Sample Plan
Sampling Method
Sampling System/Scheme
Sampling Techniques
SAP ABAP/4
SCAMPER
Scatter Diagrams
Scenario Writing
Scientific Management
Screen Design
Screening Inspection
Security and Access Controls Design
Selection Matrix
Selection Window
Self-Assessment
Self-Control
Self-Explaining Work Processes
Self-Managed Work Teams
Self-Management Enablement Skills
Self-Recording Products
Semantic Intuition
Sensitivity Analysis
Sensitivity Sampling
Sequential Analysis
Sequential Probability Ratio
Sequential Sampling Plans
Serial Equity
Servability/Serviceability
Set-Up Time Reduction
Seven Basic Tools
Shared Meeting Wall
Shelf Life
Shewhart PDCA Cycle
Ship-to-Stock Cost
Short Run Charts
Should-Cost Estimates
Sigma
Sigma Conversion Table
Signal-to-Noise Ratio
Significance Test
Simple Correlation
Simple English
Simple Language Analysis
Simple Linear Regression
Simplex-Lattice Designs
Simplification Analysis (Process, Paperwork and Language)
Simplification Approaches
Simulation Equations
Simulation Modeling
Simulation Techniques
Single Minute Exchange of Die (SMED)
Single Specification Limit Plans
Situation Analysis Diagrams
Situational Feed-Forward Diagram
Situational Feedback Diagram
Six Sigma
Six Sigma Metrics
Six Sigma System
Six-Step Error-Prevention Cycle
Six-Step Problem-Solving Cycle
Six-Step Solution-Identification Cycle
Skeletal Action Diagram
Skill-Based Compensation
Skip Lot Inspection
Skip-Lot Sampling Plans
Skunk Works
Slope (of a variable on a graph)
SMOOTH Function
Snake Chart

Social Computation
Social Connectionism
Social Expert Systems
Social Interaction Patterns
Social Process Balancing
Social Process Model
Social Responsibility
Sociogram
Sociotechnical Design
Software Configuration Management Version
Software Quality Assurance
Solace System
Solicitation Planning
Solution Analysis Diagrams
Solution Matrix
Solutions Evaluation
Source Inspection
Sources of Variability
Spearman Rank Correlation Coefficient
Specification
Specification Limits and Tolerances
Spider Diagram
Sponsor Commitment Evaluation
Sponsor Evaluation
Staffless Organizations
Stakeholder Association Matrix
Stakeholder Identification Analysis (SIA)
Stakeholder Mapping
Stakeholder Needs Analysis
Stakeholders
Standard Assessment Procedure (SAP)
Standard Deviation
Standard Rate of Work
Standardization
Standards; also see Quality Standard
Starbursting
Statistical Design of Experiments
Statistical Estimation
Statistical Inference
Statistical Methods (Control Charts)
Statistical Process Control (SPC)
Statistical Thinking
Statistical Tolerance
Statistical Tolerance Intervals/ Levels/Limits
Steepest Ascent/Descent
Stem-and-Leaf Display
Sticking Dots
Stimulus Analysis
Stock Purchase Plans
Storyboarding
Strategic Alliances Planning
Strategic Business Review
Strategic Improvement Plan
Strategic Information Systems Plan (SISP)
Strategic Plan
Stratification
Stratified Sampling
Stratum Chart
Streamlined Process Improvement (SPI)
Strengths, Weaknesses, Opportunities, and Threats Analysis (SWOT)
Stress Testing
Structural Methodology for Process Improvement
Structural Reading

Structural Roadblocks
Structure Charts
Structured Analysis/Design (SASD)
Structured Customer Surveys
Structured Design
Structured Interview
Structured Walk-Through
Student's T Distribution
Suggestion Programs
Supplier Controls
Supplier Design Involvement
Supplier, Inputs, Process, Outputs, Customers (SIPOC) Diagrams
Supplier Management
Supplier Partnerships
Supplier Process Audits
Supplier Qualification
Supplier Quality Incentive Plans
Supplier Ratings
Supplier Seminars
Supplier Surveys
Supply Chain Management
Supply Chain Technology
Surveying
SWOT Analysis
Symbolic Flowchart
Synchronous Flow Manufacturing
Synergy Survey
System
System Analysis Diagram
System Diagram
System Impact Analysis
Systematic Design
Systematic Sampling
Systems Assurance
TABLE Function
Tactical IS Plan
Taguchi's Robust Concepts
Taguchi Techniques
Takt Time
Target Costing
Target Goal Setting
Target Resistance Evaluation
Task Analysis
Task Team (TT)
Team Building
Team Management
Team Meeting Evaluation
Team Mirror
Team Process Assessment
Team Recognition
Team Resilience Questionnaire
Teams-Group Process
Technical Vitality
Technology Enablement Sessions
Technology Impact Analysis
Test Equipment
Test Objectives Definition
Test of Hypotheses
Test Plan Design
Test to Failure
Tests for Means, Variances and Proportions
Thematic Content Analysis
Theoretical Quantile
Theory of Constraints
Third-Order Delay
Three Dimensional Bar Plots
Three-Factor, Three Level Experiment
Three-Job Week
Three-Year Improvement Plan
Through Put Yield (TPY)
Tiger Teams

Time Box
Time Management
Time-Scaled Network Diagram
Time Series Analysis
Time Study Sheet
Timeline Chart
Tolerance Intervals
Tolerance Limits; Tolerances
Tollgates
Top-Down Flowchart
Top-Down Testing
Total Business Management (TBM)
Total Cost Management
Total Improvement Management (TIM)
Total HR Management (THRM)
Total Productivity Maintenance
Total Productivity Management
Total Quality Control
Total Quality Management (TQM)
Total Resource Management (TRM)
Total Six-Sigma System
Total Strategic Quality
Total Technology Management (TTM)
Toyota Production System
Traceability
Training Evaluations
Training Programs
Transition Enablers and Barriers Assessment
Transition Tree
Tree Diagrams
Trend Analysis
Triple Ranking
TRIZ
Truth Table
Two-Dimensional Survey Grid
Two-Directional Bar Chart
Two-Stage Sampling
Type I, II, III Conflicts
Type I Error Probability
Type II Error Probability
Types of Data
Types of Teams
u charts
Ubiquitous Computing
Undesirable Behavior
User Involvement
User's Quality Cost
V-Mask
Value-Added Analysis
Value-Added Process Career
Value Added to Nonvalue Added Lead Time Ratio
Value Analysis
Value Analysis and Control
Value Analysis Engineering
Value Chain Model
Value Conversation
Value Engineering
Value/Nonvalue-Added Activities
Value/Nonvalue Added Cycle Time Chart
Value Orientation
Value Propositions
Value Sensitivity Analysis
Value Stream Analysis
Value Stream Costing
Value Stream Mapping
Variability
Variable Control Charts
Variables Data
Variable Style Management

Variance Analysis
Variation Analysis
Vendor Appraisal/Supplier Evaluation
Venn Diagram
Venture Business Consult
Venture Development Workshop

Appendix C

TOOLS/METHODOLOGIES INTERACTION BETWEEN KPDS

Possible Tools *Customer Partnerships*	Management Leadership	Customer Partnerships	Business Processes	Product Processes	Service Processes	Supplier Partnerships	Improvement Mgmt.
Quality Function Deployment		■	■	■	■		
Customer Surveys		■	○	○	○		
Employee Visits		■	○	○	○		
Customer Visits		■	○	○	○		
Customer-Related Measurements		■	○	○	○		
Customer Round Tables		■	○	○	○		
Customer Interface		■	○	○	○		
Customer Phone Calls-Mgmt/Employees		■	○	○	○		
Competitive Benchmarking		■	○	○	○		
Improvement Training	■	■	■	■	■	■	
Performance Measurements		■	○	○	○	○	
Market Share Analysis		■	○	○	○		
Customer Satisfaction Measurements		■	○	○	○		
Customer Focus Groups		■	○	○	○		

■ = Should be planned for in this key change focus area
○ = Can affect this key change focus area

Possible Tools *Upper Management Involvement*	Management Leadership	Customer Partnerships	Business Processes	Product Processes	Service Processes	Supplier Partnerships	Improvement Mgmt.
Executive Improvement Team	■						■
Quality Policy Deployment	■						○
Customer Surveys		■	○	○			
Organizing for Improvement	○						■
Poor-Quality Costs		○	○	○	○		■
Management Audits			○	○	○		■
Employee Surveys	○		○	○	○		■
Communication Planning	○	○	○	○	○	○	■
Flattening the Organization	○						■
Improvement Needs Assessment	○	○	○	○	○	○	■
Improvement Czar	○	○	○	○	○	○	■
Performance Standards	■	○	○	○	○	○	○
Quality Policy	○	○	○	○	○	○	■
Quality Directives	○	○	○	○	○	○	■
No-Layoff Policy	○	○	○	○	○	○	■
Basic Improvement Rules	○	○	○	○	○	○	■
Measurement Systems	■	■	■	■	■	■	■
Improvement Funding	○	○	○	○	○	○	■
Resolve System Problems	■	○	○	○	○	○	
Organized Labor Involvement	■		○	○	○		○
New Job Descriptions	■	○	○	○	○		
Quality Education	■	■	■	■	■	■	■
Improvement Reserve Fund	■	○	○	○	○	○	○
Customer Satisfaction Measurement		■	○	○	○		

■ = Should be planned for in this key change focus area
○ = Can affect this key change focus area

Possible Tools *Upper Management Involvement (cont'd.)*	Management Leadership	Customer Partnerships	Business Processes	Product Processes	Service Processes	Supplier Partnerships	Improvement Mgmt.
Market Share Analysis		■	○	○	○		
Employee/Management Focus Groups	■		○	○	○		
Small Business Units	■		○	○	○		

■ = Should be planned for in this key change focus area
○ = Can affect this key change focus area

Possible Tools *Business Plan*	Management Leadership	Customer Partnerships	Business Processes	Product Processes	Service Processes	Supplier Partnerships	Improvement Mgmt.
Improvement Steering Council	○						■
Quality Policy Deployment	■						○
Annual Quality Plan	○	○	○	○	○	○	■
Competitive Benchmarking	○	■	○	○	○	○	
Organization's Mission	○	○	○	○	○	○	
Operating Principles	○	○	○	○	○	○	○
Business Objectives	○	○	○	○	○	○	○
Critical Success Factors	○	○	○	○	○	○	○
Performance Goals	○	○	○	○	○	○	○
Performance Strategy	○	○	○	○	○	○	○
Performance Tactics	○	○	○	○	○	○	○

Focus Areas

■ = Should be planned for in this key change focus area
○ = Can affect this key change focus area

Possible Tools *Improvement Visions and Planning*	**Focus Areas** Management Leadership	Customer Partnerships	Business Processes	Product Processes	Service Processes	Supplier Partnerships	Improvement Mgmt.
Visions and Strategic Planning	■	■	■	■	■	■	■
Improvement Needs Assessment	○	○	○	○	○	○	■
90-Day Action Plan	○	○	○	○	○	○	■

■ = Should be planned for in this key change focus area
○ = Can affect this key change focus area

Possible Tools *Management Participation*	**Focus Areas** Management Leadership	Customer Partnerships	Business Processes	Product Processes	Service Processes	Supplier Partnerships	Improvement Mgmt.
Executive Improvement Team	■						■
Leadership Skills and Training	■		○	○	○		
Delegation–Self-Managed Teams	■		○	○	○		
Management by Walking Around	■		○	○	○		
Customer Phone Calls-Mgmt/EEs	○	■	○	○	○		
Problem Solving	○	○	■	■	■	○	
Statistical Process Control	○	○	■	■	■	■	
Teams (MIT, PIT, DIT, EIT, QCC)	■	○	○	○	○	○	○
Management Audits	○	○	○	○	○	○	■
Employee Surveys	■		○	○			○
Communication Planning	○	○	○	○	○	○	■
Flattening the Organization	■	○	○	○	○	○	○
Improvement Needs Assessment	○	○	○	○	○	○	■
Management Improvement Education	■	○	○	○	○	○	
Self-Managed Work Teams	■		○	○	○		
Employee Empowerment	■		○	○	○		

■ = Should be planned for in this key change focus area
○ = Can affect this key change focus area

Possible Tools *Teams*	**Focus Areas** Management Leadership	Customer Partnerships	Business Processes	Product Processes	Service Processes	Supplier Partnerships	Improvement Mgmt.
Area Activity Analysis	■		○	○	○		
Employee Visits		■	○	○	○		
Internal Customers	■		○	○	○		
Task Teams	■	■	■	■	■	■	■
Employee Surveys	○		○	○	○		■
Problem Solving	○	○	■	■	■	■	○
Department Improvement Teams (DIT)	■	○	○	○	○	○	
Quality Circles			■	■	■		
Process Improvement Teams (PIT)		○	■	■	■	○	
Task Focus			○	○	○	○	
Business Status Reports	■		○	○	○		○
Daily Meetings	■		○	○	○		
Self-Managed Work Teams	■		○	○	○		
Team Skills Training	■	○	○	○	○	○	○
7 Basic Tools	■		○	○	○		
7 Advanced Tools	■		○	○	○		
Opportunity Cycle	■	○	○	○	○	○	
Improvement Training	○	○	○	○	○	○	■
Q&P Measurements	○	○	■	■	■	○	

■ = Should be planned for in this key change focus area
○ = Can affect this key change focus area

Possible Tools *Individual Involvement*	**Focus Areas:** Management Leadership	Customer Partnerships	Business Processes	Product Processes	Service Processes	Supplier Partnerships	Improvement Mgmt.
Performance Planning & Appraisal	■		○	○	○		
Suggestion Systems	■	○	○	○	○	○	○
Measurement and Feedback	■	■	■	■	■	■	■
Employee Visits		■	○	○	○		
Customer Interface Training		■	○	○	○		
Internal Customer	■		○	○	○		
Job Training and Certification	■	■	■	■	■	■	
Problem Solving	■	○	○	○	○	○	○
Pay for Performance	○	○	○	○	○		■
Career Planning	■		○	○	○		
Job Improvement Recognition Program	○	○	○	○	○	○	■
Request for Corrective Action			○	○	○		■

■ = Should be planned for in this key change focus area
○ = Can affect this key change focus area

Possible Tools *Process Development*	Management Leadership	Customer Partnerships	Business Processes	Product Processes	Service Processes	Supplier Partnerships	Improvement Mgmt.
Measurement and Feedback	■	■	■	■	■	■	■
Quality Function Deployment		■	■	■	■		
Customer Surveys		■	○	○	○		
Customer Visits		■	○	○	○		
Customer Interface Training		■	○	○	○		
Customer Phone Calls-Mgmt/Employees		■	○	○	○		
Process Benchmarking		■	■	■	■	■	
Business Process Improvement	○	○	■	○	■	○	
Bureaucracy Elimination	○	○	■	○	■	○	
Internal Customer	■		○	○	○		
Process Improvement Teams	○	○	■	■	■	○	○
Task Teams	○	○	■	■	■	○	○
Design Review				■		■	
Manufacturability Assessment				■			
Job Training and Certification	○	○	■	■	■	○	
Production Process Improvement		○		■		○	
Problem Solving	○	○	■	■	■	■	
Statistical Process Control	○	○	■	■	■	■	
Design of Experiments		○	■	■	■	○	
Standardization	○	○	■	■	■		
Zero Stock/Just-in-Time			○	■	○	■	
Failure Analysis		○		■	○	■	
Product Cycle Controls		○	○	■	■	○	
Reliability-Centered Maintenance		○		■			

■ = Should be planned for in this key change focus area
○ = Can affect this key change focus area

Possible Tools *Process Development (cont'd.)*	**Focus Areas** Management Leadership	Customer Partnerships	Business Processes	Product Processes	Service Processes	Supplier Partnerships	Improvement Mgmt.
Cost Focus, Not Price	○	■	■	■	■	■	
Poor-Quality Cost		○	○	○	○		■
Process Qualification		○	■	■	■	■	
Improvement Needs Assessment	○	○	○	○	○	○	■
Concurrent Engineering			○	■	■	■	
Error Proofing	○	○	■	■	■	○	
Automation		○	■	■	■	○	

■ = Should be planned for in this key change focus area
○ = Can affect this key change focus area

Possible Tools *Supplier Partnerships*	**Focus Areas** Management Leadership	Customer Partnerships	Business Processes	Product Processes	Service Processes	Supplier Partnerships	Improvement Mgmt.
Task Teams	■	■	■	■	■	■	■
Problem Solving	○	○	■	■	■	○	
Design of Experiments		○	■	■	■	○	
Standardization	○	○	■	■	■	○	
Zero Stock/Just-in-Time			○	■	○	■	
Failure Analysis		○		■	○	■	
Supplier Surveys						■	
Supplier Seminars						■	
Supplier Incentive Plans						■	
Cost Focus, Not Price				○	○	■	
Supplier Qualification				○		■	
Concurrent Engineering				■		■	
ISO 9000	○	○	○	■	○	■	
Improvement Needs Assessment	○	○	○	○	○	○	■
Fewer Suppliers						■	
Long-Term Contracts						■	
Statistical Process Control	○	○	■	■	■	■	
Source Audits						■	
Supplier Ratings						■	
Supplier Recognition						■	

■ = Should be planned for in this key change focus area
○ = Can affect this key change focus area

Possible Tools *Systems Assurance*	Management Leadership	Customer Partnerships	Business Processes	Product Processes	Service Processes	Supplier Partnerships	Improvement Mgmt.
Quality Planning			○	○	○		■
Quality Assurance Systems			○	○	○		■
ISO 9000	○	○	○	■	○	■	
Improvement Needs Assessment	○	○	○	○	○	○	■
Reliability Analysis		○		■			
Maintainability Analysis		○		■			
Product Assurance		○		■			
Quality Laboratories		○		■			
In-Process Inspection		○	■	■	■		
Systems Audits	○	○	○	○	○	○	■
Department Audits	○	○	○	○	○	○	■

(Column headers grouped under **Focus Areas**)

■ = Should be planned for in this key change focus area
○ = Can affect this key change focus area

Possible Tools *Rewards and Recognition*	**Focus Areas** Management Leadership	Customer Partnerships	Business Processes	Product Processes	Service Processes	Supplier Partnerships	Improvement Mgmt.
Rewards and Recognition Systems	○		○	○			■
Improvement Needs Assessment	○	○	○	○	○	○	■
Bonuses	○	○	○	○	○	○	■
Gain Sharing	○	○	○	○	○	○	■
Team Bonus Plans	○	○	○	○	○	○	■
Stock Options	○	○	○	○	○	○	■
Contribution Awards	○	○	○	○	○	○	■
Pay for Performance	○	○	○	○	○	○	■
Suggestion System	■	○	○	○	○	○	○
Patent Awards	○	○	○	○	○	○	■
Individual Public Recognition	○	○	○	○	○	○	■
Group Recognition	○	○	○	○	○	○	■
Group Awards	○	○	○	○	○	○	■
Private Recognition	○	○	○	○	○	○	■
Commissions		■					

■ = Should be planned for in this key change focus area
○ = Can affect this key change focus area

Index

D

E

K

L

M

Q

R

S

W

Y

About the Author

H. James Harrington, PhD. In the book, *Tech Trending*, Dr. Harrington was referred to as "the quintessential tech trender." The *New York Times* referred to him as having a "… knack for synthesis and an open mind about packaging his knowledge and experience in new ways—characteristics that may matter more as prerequisites for new-economy success than technical wizardry…". The author, Tom Peters, stated, "I fervently hope that Harrington's readers will not only benefit from the thoroughness of his effort, but will also 'smell' the fundamental nature of the challenge for change that he mounts." Bill Clinton, former president of the United States, appointed Dr. Harrington to serve as an Ambassador of Goodwill. It has been said about him: "He writes the books that other consultants use."

Harrington Institute (Los Gatos, California) was featured on a half-hour TV program, *Heartbeat of America*, which focuses on outstanding small businesses that make America strong. The host, William Shatner, stated: "You (Dr. Harrington) manage an entrepreneurial company that moves America forward. You are obviously successful."

At present, Dr. Harrington serves as the CEO for the Harrington Institute and Harrington Middle East. He also serves as the chairman of the board for a number of businesses. Dr. Harrington is recognized as one of the world leaders in applying performance improvement methodologies to business processes. He has an excellent record of coming into an organization, working as its CEO or COO, resulting in a major improvement in its financial and quality performance.

In February 2002, Dr. Harrington retired as the COO of Systemcorp A.L.G., the leading supplier of knowledge management and project management software solutions when Systemcorp was purchased by IBM. Prior to this, he served as a principal and one of the leaders in the Process Innovation Group at Ernst & Young. He retired from Ernst & Young when it was purchased by Cap Gemini. Dr. Harrington joined Ernst & Young when that company purchased Harrington, Hurd, & Rieker, a consulting firm that Dr. Harrington started. Before that, he was with IBM for over 40 years as a senior engineer and project manager.

Dr. Harrington is past chairman and past president of the prestigious International Academy for Quality and of the American Society for Quality Control. He is also an active member of the Global Knowledge Economics Council.

H. James Harrington was elected to the honorary level of the International Academy for Quality, which is the highest level of recognition in the quality profession.

Dr. Harrington's contributions to performance improvement around the world have brought him many honors. He was appointed the honorary advisor to the China Quality Control Association, and was elected to the Singapore Productivity Hall of Fame in 1990. He has been named lifetime honorary president of the Asia-Pacific Quality Control Organization and honorary director of the Association Chilean de Control de Calidad.

He has been elected a fellow of the British Quality Control Organization and the American Society for Quality Control. In 2008, he was elected to be an honorary fellow of the Iran Quality Association and Azerbaijan Quality Association. He also was elected an honorary member of the quality societies in Taiwan, Argentina, Brazil, Colombia, and Singapore. He is listed in the *Who's Who Worldwide* and *Men of Distinction Worldwide.* He has presented hundreds of papers on performance improvement and organizational management structure at the local, state, national, and international levels.

Other recognitions include:

- The Harrington/Ishikawa Medal, presented yearly by the Asian Pacific Quality Organization, was named after H. James Harrington to recognize his many contributions to the region.
- The Harrington/Neron Medal was named after H. James Harrington in 1997 for his many contributions to the quality movement in Canada.
- Harrington Best TQM Thesis Award was established in 2004 and named after H. James Harrington by the European Universities Network and e-TQM College.
- Harrington Chair in Performance Excellence was established in 2005 at the Sudan University.
- Harrington Excellence Medal was established in 2007 to recognize an individual who uses the quality tools in a superior manner.
- H. James Harrington Scholarship was established in 2011 by the ASQ Inspection Division.

H. James Harrington has received many awards and recognition throughout the years, which include:

- 1978: Benjamin L. Lubelsky Award
- 1982: The John Delbert Award
- 1980: The Administrative Applications Division Silver Anniversary Award
- 1996: The ASQC's Lancaster Award
- 2001: The Magnolia Award in recognition for the many contributions he has made in improving quality in China
- 2002: Selected by the European Literati Club to receive a lifetime achievement award
- 2002: The International Academy of Quality President's Award
- 2003: The Edwards Medal from the American Society for Quality (ASQ).
- 2004: The Distinguished Service Award which is ASQ's highest award for service granted by the Society
- 2008: The Sheikh Khalifa Excellence Award (UAE) in recognition of his superior performance as an original Quality and Excellence Guru who helped shape modern quality thinking
- 2009: The Professional of the Year
- 2009: The Hamdan Bin Mohammed e-University Medal
- 2010: The Asian Pacific Quality Association (APQO) President's Award
- 2010: The Australian Organization of Quality NSW's Board recognized Harrington as "the Global Leader in Performance Improvement Initiatives"
- 2011: The Shanghai Magnolia Special Contributions Award from the Shanghai Association for Quality in recognition of his 25 years of contributing to the advancement of quality in China
- 2012: ASQ Ishikawa Medal
- 2012: The Jack Grayson Award. This award recognizes individuals who have demonstrated outstanding leadership in the application of quality philosophy
- 2012: The A.C. Rosander Award, which is ASQ Service Quality Division's highest honor.
- 2012: The Armand V. Feigenbaum Lifetime Achievement Medal by the Asia Pacific Quality Organization

Dr. Harrington is a very prolific author, publishing hundreds of technical reports and magazine articles. For eight years, he was published a monthly column in *Quality Digest Magazine* and is syndicated in five other publications. He has authored 39 books and 10 software packages. You may contact Dr. Harrington at:

16080 Camino del Cerro, Los Gatos, California, 95032.
Phone: (408) 358-2476
hjh@harrington-institute.com.

Made in the USA
Coppell, TX
01 February 2022